The Pig

by Sanders Spencer

CONTENTS

INTRODUCTION

There are few points in the breeding of stock on which a greater variation of opinion has been confidently expressed than on the origin of the domesticated pig. It has been contended that our various types had a common origin in the wild hog, and that the difference in form, colour, and character amongst the local breeds is due, in the main, to the requirements, imaginary or real, of the interested residents in the particular districts. On the other hand, it is asserted with equal confidence, and probably with the same amount of actual proof, that it would be impossible so to improve the wild hog by selection as to render it the equal of the domesticated hog. There must, therefore, have been an infusion of blood of a cultivated breed of pigs to acquire even that amount of success which was noticeable in the improved pig of a century, or less, ago. Unfortunately, for this argument, it has not been possible to obtain any information of value as to the alleged source of origin of this cultivated breed of pigs.

Again, those pigs which possess in a marked degree early maturity, fine quality of flesh, and those other characteristics of the improved pig, are so various in colour, that one cultivated breed only could not have been utilised in the general improvement.

All the ancient writers on pigs appear to have experienced the same difficulty when endeavouring to discover the sources of origin of the material which might have been used in the production of the pig which in their time was looked upon as the domesticated and improved pig of the period. This difficulty extends even to the present day. So far as is known there exists no actual proof that the domesticated hog has been evolved in any particular way other than by continued selection of those animals for breeding purposes which possessed in the greatest degree those particular qualities held in the highest estimation at the time.

Of course, the soil, climate, etc., of the district in which pigs are reared have a certain amount of influence, but this is noticeable to a much lesser extent with pigs than with horses, cattle, or sheep, since under the present system of pig-breeding the greater portion of the food used in the different districts is of a very similar character--indeed, much of it has a common origin--having been imported from abroad.

As a rule, comparatively speaking very little difference is noticeable in the development, form, and character of pigs bred in the various parts of the country, whereas with some of the other domesticated animals a very considerable change follows the removal of sheep of a pure breed from one district to another. The quality and quantity of the wool, flesh, and bone are all affected. An exactly similar effect is noticeable when horses of a particular breed are moved from one district to another. For instance, a Shire foal bred in the Fens may possess the characteristic bone, flesh, and hair, yet if it be moved into portions of the Eastern counties where the soil is of a totally different character, it will when matured have lost, to a very considerable extent, its peculiar characteristics of bone and hair. The changes wrought may be due in small measure to climate, but the predominant cause must be due to the variation in the food grown on soils of a different character.

This question of the original cause or causes of the varying colour of the pigs in different localities appears to be equally difficult of solution. As to the continuation in certain districts of pigs of one colour, custom and even prejudice have a great effect. So strong is this prejudice that some persons will even declare that the pork of pigs of the fashionable colour in the neighbourhood is superior to that from pigs of any other colour. As this weakness is common in districts where black and where white pigs are kept it must be admitted that prejudice alone must be the foundation of the belief.

Probably the safest conclusion to arrive at with respect to the variation in colour of the pigs noticeable in certain districts is that in the long ago the native pig in the wild state was of the colour of the soil and the herbage In which it sheltered, and was thus less conspicuous to its enemies, whether human or animal. A marked instance of this is to be found in the colour of the common or original pig found in some parts of the country where the soil is of a decidedly red colour. In the district referred to one actually hears some farms spoken of as "red land farms." What more natural than to find in the districts in which land of this hue predominates that the pigs should be a red rusty hue such as was the original colour of that breed of pigs now called the Tamworth breed.

Some persons, who do not agree with this theory of the origin of the various coloured pigs, cite as a proof of their belief the fact that in so many districts the pigs are of a mixed colour, and that this peculiar marking is equally as general in localities as is any particular or special colour in the pigs. This is perfectly true, and it is probably due to exactly the same causes, fancy, prejudice, or custom. The residents in certain districts have grown accustomed to certain things or certain forms, and are loth to change; the manufacturer of any article must humour the actual or fancied requirements of his customers If he is to secure success; and in a similar manner the breeder of pigs has to consider and to produce pigs of the form, size, and colour which are most in demand. Further if, as confidently alleged, there is a preference in some districts for pork from pigs of a certain colour, then the butcher naturally offers a higher price for pigs of that colour which most fully satisfy the fancies of his customers, and thus we find a similarity of form and colour in the pigs of various districts.

As to the origin of these parti-coloured pigs, the explanation offered is that even in pre-railway times there was a certain amount of interchange of the different local breeds of stock. This would be affected in various ways, which need not be specified.

At the present time we have several defined and distinct breeds of pigs which have secured recognition at our principal agricultural shows. Indeed it may be claimed that the exhibitions of live stock which have become so general in all parts of the country have been one of the chief factors in fixing to a certain extent the type and character of certain local breeds. Within the memory of the present writer the classification of pigs at our principal shows was of a very simple character; it consisted of classes for pigs of a white colour and for pigs of any other colour. There was no attempt at any definition as to size, form, and quality of the pigs. These points were left entirely to the judges, who naturally were led to favour pigs of the type which they bred. There was thus a greater amount of uncertainty as to the success of an exhibitor's stock than at the present time. This uncertainty--save as to the members of the Show Committees or their friends--was increased by the unfair system of withholding from the knowledge of the average exhibitor the names of those selected to judge.

The necessity of some definition, if only of colour, quickly became obvious. At first classes were established for pigs of certain colours; then the prizes were offered for pigs of certain breeds, which were more or less loosely defined. Now at the chief shows the pigs exhibited in the various classes must be qualified for entry in the herd books of the particular breeds.

At the Royal Agricultural Shows there have been classes for pigs of the Large White, Middle White, Berkshire, Tamworth, Large Black and Lincolnshire Curly Coated breeds; whilst for the next show classes for pigs of the so-called Gloucestershire Old Spots breed are to be included. As showing the changes which are in progress it may be noted that two breeds of pigs which had classes provided for them at the Royal and some other Shows have become extinct. These were the Small White and the Small Black breeds--the sole cause of their disappearance being the unsuitability of the pigs of the breeds to supply the present requirements of the consumer.

* * * * *

THE PIG

CHAPTER I

NON-PEDIGREE PIGS

Although the more general use of so-called pedigree pigs has tended to modify the characteristics of the various local breeds of pigs, yet it is possible to find a certain number of pig breeders who adhere to the type of pig which has been in the past most generally found in their district. This type was undoubtedly fixed by the wants or fancies of those resident in the particular portions of the country.

In the past it has been the practice when describing these local breeds to write as though they were confined to certain counties. It may be that pigs of a peculiar or characteristic type are more numerous within the borders of various counties, but this is by no means always the case. The habits and pursuits of the inhabitants rather than the soil and climate--as with horses, cattle and sheep--have the greatest influence on the form, size and quality of the local pigs, whilst use and custom appear to determine the colour of the pig. We are of opinion that it will be more instructive if we give a short description of some of the more common types of these local breeds of pigs, and mention the names of those counties in which they are more generally found.

Amongst the most distinct of these local breeds is that which is variously termed the sheeted or saddle-backed pig, which in the United States has a society to look after its interests, and where it bears the name of

THE HAMPSHIRE

Just why our American cousins should have decided to call these sheeted pigs Hampshires is not on the surface, since the oldest writers on pigs give to the county of Sussex the credit of being their original home. The description given by Sidney of the Hampshire pig is that "it is a coarse and useful black pig, inferior to the Berkshire, and not in the same refined class as the Essex." Richardson writes "The Hampshire breed is not infrequently confounded with the Berkshire; but its body is longer and its sides flatter; the head is long and the snout sharp. The colour of the breed is usually dark spotted; but it is sometimes black altogether, and more frequently white."

The sheeted pig has also been bred in the county of Essex for over a century, but it is recorded that it was introduced into this county by a Mr. Western who subsequently became Lord Western, and whose estate was situated in Essex.

In Sidney's book The Pig, we read, "West Sussex, Hampshire, Berkshire, Dorset, Shropshire and Wales had indigenous black or red and black breeds of swine; and between the whites, the blacks and the reds the parti-colours were produced which have since in a great degree disappeared under the influence of prizes, generally awarded to pure breeds of single colours."

Sidney also states "that Youatt and all the authors who have followed him down to the latest work published on the subject, occupy space in describing various county pigs which have long ceased to possess, if they ever possessed, any merit worth the attention of the breeder."

The Rudgwick, which is another name for the sheeted pig, is included in the list. Richardson at a still earlier date describes the Sussex breed as "black and white in colour, but not spotted; that is to say, these colours are distributed in very large patches; one half--say, for instance, the fore part of the body white and the hinder end black; or sometimes both ends black and the middle white or vice vers? these pigs are in no way remarkable; they seldom feed to over twenty stone. They are well made, of middle size, and their skin covered with scanty bristles. The snout tapering and firm, the ears upright and pointed, the jowl deep and the body compactly round. They arrive at early maturity, fatten quickly, and the flesh is excellent."

Richardson also writes, "There is another improved Essex breed, called the Essex Half-Blacks, resembling that which I have described in colour, said to be descended from the Berkshire. This breed was originally introduced by Lord Western, and obtained much celebrity," etc. etc.

He then quotes from The Complete Grazier, sixth edition, as follows: "They are black and white, short haired, fine skinned with smaller heads and ears than the Berkshire, but feathered with inside hair which is a distinctive mark of both; have short snubby noses, very fine bone, broad and deep in the belly, full in the hind quarters, but light in the bone and offal. They feed remarkably fast and are of an excellent quality of meat. The sows are good breeders, and bring litters of from eight to twelve; but they have the character of being bad nurses." If this allegation were true at the time it was written, it is not at the present time as the Half-Blacks or sheeted sows are both prolific and first-rate mothers.

Malden describes the Sussex: "A large breed called the Rudgwick, was one of the largest in England. There appears to be a doubt as to whether the coloured pig was descended from the spotted Berkshire or the black and white Essex. They were of medium size, of good quality generally, but of somewhat heavy bone." The generally accepted view is that the Essex sheeted pig was descended from importations from the county of Sussex. These sheeted pigs are still occasionally met with in Essex, but the system of crossing which is generally followed by pig breeders in the county is gradually reducing its number, although even amongst the cross-breeds the peculiar marking occasionally shows itself. At the time of writing there is a movement on foot to form a society for the purpose of reviving the breed. From the utility point of view the sheeted pig has much to recommend it, but whether or not one or more of the breeds of pigs whose pedigrees are already recorded do not possess at least equal merits must be left for decision by others.

SPOTTED PIGS

In many districts are found other parti-coloured pigs, but in these the black, the white, and the red colours show themselves in spots of varying size and extent. Probably amongst the best types of these spotted pigs is the one found over the greater part of the county of Northampton, and portions of the counties of Leicester and Oxford adjoining. In the former county the pigs have more of black than white in their colouring, whilst in the two latter red spots are more often seen. This is probably due to a stronger infusion of the blood of the Staffordshire red pig which is now known as the Tamworth. The blood of the Neapolitan pig through the Berkshire or the Small Black is credited with being the origin of the darker coloured Northamptonshire spotted pig. The qualities claimed for these spotted or "plum pudding" pigs as they are locally termed, are prolificacy, quick growth, hardihood, and the production of pork possessing a large proportion of lean to fat meat. They are also good grazers, and grow to a size quite the equal of the Berkshire. In form they are perhaps more suited for the fresh pork trade than for the manufacture of bacon of the kind now so much in demand.

THE LARGE WHITE AND BLUE PIGS

Those large, coarse-boned pigs with hair of a white colour and skins more or less mottled with blue are gradually giving place to pigs with finer hair, skin, bone, and quality of meat. The coarse lop ears are being reduced in size and thickness, whilst the pig itself is becoming less gaunt and its early maturity considerably increased by crossing with the better quality Large White and the quickly maturing Middle White. These coarse white with blue markings pigs were common in the Fens of Cambridgeshire, Norfolk, the Isle of Ely and Lincolnshire, and in the counties of Bedford, Cheshire, etc.

WHITE PIGS

Within the memory of persons now living, white pigs of varying types were found in various parts of this country. Many of these white pigs found in Norfolk, Suffolk, Shropshire, and Wales had little to recommend them as they were flat sided, long legged, hard feeders, and required to be comparatively old before they could be turned into pork. A vast improvement has of late years been effected in these unprofitable swine by crossing them with compact and early maturing pigs of different colours, but mainly white pigs until the last few years, when Large Blacks and even a few Gloucestershire Old Spots boars have been introduced in Norfolk.

At one time white pigs of a small size were by no means uncommon in Suffolk, Essex, Middlesex, Yorkshire, and parts of Berkshire, and other counties. The origin of these small, compact, and early maturing pigs appears to have been a cross of the imported Chinese on the neater and shorter country pigs of a white colour. For a period these handsome pigs were quite fashionable amongst the well-to-do, but the general public objected to the pork produced by them, owing to its excessive fatness. The bacon curers still more strongly objected to the short sides and the very small amount of lean meat in the cured carcases. During the last thirty years comparatively few of these pretty, but useless, pigs have been bred.

BLACK PIGS

The description given of the two main types of white pigs would apply equally well to the Black pigs common in this country, save with respect to colour. The long flat-sided black pig was found in Essex, Suffolk, Cambridgeshire, Sussex, etc. These pigs were noted for their prolificacy, hardihood, and quick growth, whilst the sows furnished a full supply of milk to their youngsters, but they were such slow feeders that it became necessary to cross them with pigs which matured more quickly. A type of black pig similar in form to the Small White was also found in Essex and Suffolk, whilst in Devonshire, Dorset, and one or two other counties the colour of the pigs was blue rather than black, and of a somewhat larger size, but possessing the same weakness, too large a proportion of fat to lean meat.

CHAPTER II

PURE BREEDS

The task of writing a description of the various breeds of swine has been rendered less difficult by the formation during the past half-century of societies for the registration of the pedigrees of the pigs of the different breeds, and by the setting up of scales of those points which pigs for entry in the particular herd books should possess. The first of these societies was the National Pig Breeders Association, of which the present writer was the honorary secretary for two years. At the time of its formation the breeds of pigs most generally recognised were the Berkshire, the Large, Middle, and Small Whites or Yorkshires, and the Small Black breed. It was intended that the pedigrees of the pure bred pigs of each of these breeds should be recorded by the Association and published in one herd book.

There is no doubt that this would have been an ideal plan, and would have resulted in a saving of much labour and expense, and decidedly more convenient for those connected with the export trade. For reasons into which it is not now necessary to enter, the breeders of Berkshires determined to have a separate herd book; therefore, they started a society which they named the British Berkshire Society, to distinguish it from the American Berkshire Record.

Subsequently the Tamworth breed of pigs became recognised by the Royal Agricultural Society, and the breeders of the red pig joined the National Pig Breeders Association. Then the demand for Small White and Small Black pigs ceased, so that eventually the pedigrees of Large White, Middle White, and Tamworth pigs only were registered in the N.P.A. Herd Book.

Subsequently the breeders of Tamworth pigs formed themselves into a society presumably for propaganda work, and to conserve the interests of breeders of Tamworth pigs. Of late years other breeds of pigs have been brought to public notice, and have had herd books, and societies specially devoted to their particular interests. The Large Black, Large White Ulster, the Lincolnshire Curly Coated pigs, the Gloucestershire Old Spots, and the Cumberland pigs have their pedigrees recorded. An attempt was made some years since to resuscitate the Oxfordshire Spotted pig, but it was not a continued success. It is quite possible that other local breeds of pigs may find sufficient admirers to form societies to bring before the public the many good qualities possessed by the pigs of these breeds, but apart from local interest it is at least doubtful if any permanent benefit will supervene from this multiplication of herd books--save that it may increase the interest in pig breeding, a result devoutly to be prayed for.

The issuing of the scales of points of those breeds of pigs whose pedigrees are recorded in the various herd books has rendered it unnecessary for us to endeavour to formulate the good qualities which are presumably those which are more or less completely possessed by these pedigreed animals, nor does the necessity exist for us to mention those particular qualities which each breed is supposed by the admirers of other breeds to lack. There is no doubt that each breed possesses certain points which render it specially suitable for differing localities and varying purposes.

Some persons who look upon a pig solely as an animal, as a converter of various substances into pork, are of opinion that the tendency of those responsible for the running of these societies is towards fancy points to the detriment of the practical points. There appears to have been some grounds for this view. The Small White, the Small Black, the Berkshire, and the Large White have all been affected by the acts of faddists. The three first-named breeds suffered from the aims of certain of the breeders to reduce the size and to increase the so-called quality until the consumers of pork refused to follow the fashion; whilst the craze which has seriously affected the utility of the Large White pigs has been exactly the opposite, i.e. an endeavour to so vastly increase the size that they ceased to supply the kind of pork and the size of joints which the general public demanded. It may be natural for fanciers to declare that a Small White or a Small Black pig must be a small animal, but this is only on comparison with the large breeds of the same colour and characteristics. The usefulness of the pig in the commercial world must be studied if any breed of pig is to hold its own on the market.

The opposite extreme to that followed by the breeders of the small breeds is that of the breeders of Large Whites, who look upon mere slze as the most important of the points to be studied. The mere increase in bulk, in length of head and leg and weight of bone may appeal to the mere fancier or faddist, but by paying undue attention to these fancy points the actual object of the breeding and fattening of pigs is lost sight of, and the consumer who is after all the one whose wants must first receive study, is estranged and the commercial market is lost.

In the following pages will be found full particulars together with the scales of points, as issued by the various societies, of the chief breeds and varieties.

* * * * *

STANDARD OF EXCELLENCE

LARGE WHITE

COLOUR.--White, free from black hairs, and as free as possible from blue spots on the skin.

HEAD.--Moderately long, face slightly dished, snout broad, not too much turned up, jowl not too heavy, wide between the ears.

EARS.--Long, thin, slightly inclined forward, and fringed with fine hair.

NECK.--Long, and proportionately full to shoulders.

CHEST.--Wide and deep.

SHOULDERS.--Level across the top, not too wide, free from coarseness.

LEGS.--Straight and well set, level with the outside of the body with flat bone.

PASTERNS.--Short and springy.

FEET.--Strong, even, and wide.

BACK.--Long, level, and wide from neck to rump.

LOIN.--Broad.

TAIL.--Set high, stout and long, but not coarse, with tassel of fine hair.

SIDES.--Deep.

RIBS.--Well sprung.

BELLY.--Full, but not flabby, with straight under line.

FLANK.--Thick, and well let down.

QUARTERS.--Long and wide.

HAMS.--Broad, full, and deep to hocks.

COAT.--Long and moderately fine.

ACTION.--Firm and free.

SKIN.--Not too thick, quite free from wrinkles.

Large bred pigs do not fully develop their points until some months old, the pig at five months often proving at a year or 15 months a much better animal than could be anticipated at the earlier age and vice vers? but size and quality are most important.

OBJECTIONS.--Black hairs, black spots, a curly coat, a coarse mane, short snout, inbent knees, hollowness at back of shoulders.

* * * * *

MIDDLE WHITE

COLOUR.--White, free from black hairs or blue spots on the skin.

HEAD.--Moderately short, face dished, snout broad and turned up, Jowl full, wide between ears.

EARS.--Fairly large, carried erect and fringed with fine hair.

NECK.--Medium length, proportionately full to the shoulders.

CHEST.--Wide and deep.

SHOULDERS.--Level across the top, moderately wide, free from coarseness.

LEGS.--Straight and well set, level with the outside of body with fine bone.

PASTERNS.--Short and springy.

FEET.--Strong, even, and wide.

BACK.--Long, level, and wide from neck to rump.

LOIN.--Broad.

TAIL.--Set high, moderately long, but not coarse, with tassel of fine hair.

SIDES.--Deep.

RIBS.--Well sprung.

BELLY.--Full, but not flabby, with straight under line.

FLANK.--Thick and well let down.

QUARTERS.--Long and wide.

HAMS.--Broad, full, and deep to hocks.

COAT.--Long, fine, and silky.

ACTION.--Firm and free.

SKIN.--Fine, and quite free from wrinkles.

OBJECTIONS.--Black hairs, black or blue spots, a coarse mane, inbent knees, hollowness at back of shoulders, wrinkled skin.

* * * * *

TAMWORTH

COLOUR.--Golden red hair on a flesh coloured skin, free from black.

HEAD.--Fairly long, snout moderately long and quite straight, face slightly dished, wide between ears.

EARS.--Rather large, with fine fringe, carried rigid and inclined slightly forward.

NECK.--Fairly long and muscular, especially in boar.

CHEST.--Wide and deep.

SHOULDERS.--Fine, slanting, and well set.

LEGS.--Strong and shapely, with plenty of bone and set well outside body.

PASTERNS.--Strong and sloping.

FEET.--Strong, and of fair size.

BACK.--Long and straight.

LOIN.--Strong and broad.

TAIL.--Set on high and well tasselled.

SIDES.--Long and deep.

RIBS.--Well sprung and extending well up to flank.

BELLY.--Deep, with straight under line.

FLANK.--Full and well let down.

QUARTERS.--Long, wide, and straight from hip to tail.

HAMS.--Broad, and full, well let down to hocks.

COAT.--Abundant, long, straight, and fine.

ACTION.--Firm and free.

OBJECTIONS.--Black hair, very light or ginger hair, curly coat, coarse mane, black spots on skin, slouch or drooping ears, short or turned up snout, heavy shoulders, wrinkled skin, inbent knees, hollowness at back of shoulders.

* * * * *

BERKSHIRE PIGS

COLOUR.--Black, with white on face, feet and tip of tail.

SKIN.--Fine, and free from wrinkles.

HAIR.--Long, fine, and plentiful.

HEAD.--Moderately short, face dished, snout broad; and wide between the eyes and ears.

EARS.--Fairly large, carried erect or slightly inclined forward, and fringed with fine hair.

NECK.--Medium length, evenly set on shoulders; jowl full and not heavy.

SHOULDERS.--Fine and well sloped backwards; free from coarseness.

BACK.--Long and straight, ribs well sprung, sides deep.

HAMS.--Wide and deep to hocks.

TAIL.--Set high, and fairly large.

FLANK.--Deep and well let down, and making straight under line.

LEGS AND FEET.--Short, straight, and strong, set wide apart, and hoofs nearly erect.

IMPERFECTIONS.--A perfectly black face, foot, or tail. A white ear. A crooked jaw. White or sandy spots, or white skin on the body. A rose back. A very coarse mane, and inbent knees.

* * * * *

LARGE BLACK PIG

SCALE OF POINTS

HEAD.--Medium length and wide between the ears 5

EARS.--Thin, inclined well over the face, and not extending beyond point of nose 4

JOWL.--Medium size 3

NECK.--Fairly long and muscular 3

CHEST.--Wide and deep 3

SHOULDERS.--Well developed, in line with the ribs 8

BACK.--Long and level 15

RIBS.--Well sprung 5

SIDES.--Very deep 8

LOIN.--Broad 5

BELLY AND FLANK.--Thick and well developed 7

QUARTERS.--Long, wide, and not drooping 8

HAMS.--Large and well filled to hocks 10

TAIL.--Set high, of moderate size 3

LEGS.--Short, straight, flat, and strong 5

SKIN AND COAT.--Fine and soft, with moderate quantity of straight, silky hair 8

--- 100

OBJECTIONS.--Head--narrow forehead or dished nose. Ears--thick, coarse, or pricked. Coat--curly or coarse, with rose, bristly mane. Skin--wrinkled.

DISQUALIFICATION.--Colour--any other than black.

* * * * *

LARGE WHITE ULSTER

SCALE OF POINTS

HEAD.--Moderately long, wide between the ears 5

EARS.--Long, thin, and inclined well over the face 6

JOWL.--Light 5

NECK.--Fairly long and muscular 2

CHEST.--Wide and deep 3

SHOULDERS.--Not coarse, oblique, narrow plate 8

LEGS.--Short, straight, and well set, level with the outside of the body, with flat bone, not coarse 5

PASTERNS.--Straight 5

BACK.--Long and level (rising a little to centre of back not objected to) 12

SIDES.--Very deep 10

RIBS.--Well sprung 5

LOIN.--Broad 3

QUARTERS.--Long, wide, and not drooping 8

HAMS.--Large and well filled to hocks 12

BELLY AND FLANK.--Thick and well filled 5

TAIL.--Well set and not coarse 1

SKIN.--Fine and soft 10

COAT.--Small quantity of fine silky hair 10

--- Total 100

OBJECTIONS.--Head--narrow forehead. Ears--thick, coarse, or pricked. Coat--coarse or curly; bristly mane.

DISQUALIFICATION.--Colour--any other than white.

* * * * *

LINCOLNSHIRE CURLY-COATED PIG

SCALE OF POINTS

COLOUR.--White

FACE AND NECK.--Medium length and wide between the eyes and ears 5

EARS.--Medium length, and not too much over face 10

JOWL.--Heavy 3

CHEST.--Wide and deep 3

SHOULDERS.--Wide 15

BACK.--Long and level 10

SIDES.--Very deep, and ribs well sprung 10

LOIN.--Broad 5

QUARTERS.--Long, wide, and not drooping 5

HAMS.--Large and well filled to hocks 15

TAIL.--Set high and thick 3

LEGS.--Short and straight 5

BELLY AND FLANK.--Thick and well filled 3

COAT.--Fair quantity of curly or wavy hair 8

--- 100

OBJECTIONS.--Head--narrow forehead. Ears--Thin.

DISQUALIFICATIONS.--Ears--pricked. Nose--dished or long. Coat--coarse, straight, or bristly. Colour of hair--any other than white.

* * * * *

THE GLOUCESTERSHIRE OLD SPOTS

HEAD.--Medium length and wide between the ears, nose wide and medium length, slightly dished.

EARS.--Rather long and drooping.

JOWL.--Medium size.

NECK.--Fairly long and muscular.

CHEST.--Wide and deep.

SHOULDERS.--Well developed but not projecting and in line with ribs, must not show any coarseness.

BACK.--Long and level.

RIBS.--Deep, well sprung.

LOIN.--Very broad.

SIDES.--Very deep and presenting straight bottom line.

BELLY AND FLANK.--Full and thick.

QUARTERS.--Long, wide, and not drooping.

TAIL.--Set high, of moderate size, yet fairly strong and long and carrying brush.

HAMS.--Large, not too flat, and well filled to the hocks.

LEGS.--Short, straight and strong.

SKIN AND COAT.--Skin light or dark, must not show coloured splotches otherwise than beneath the spots of the coat. The latter should be full and fairly thick, hair long and silky but not curly, with an absence of mane bristles. Colour: white spots on black ground, or black spots on white ground. Such spots to be of medium size.

TEATS.--Minimum number of teats to be considered.

OBJECTIONS.--Head--narrow, face and nose both dished. Ears--thick, floppy, coarse, or elevated. Coat--Coarse or curly with rose; bristly mane, or decidedly sandy colour; skewbald or saddleback markings.

* * * * *

THE CUMBERLAND PIG

HEAD.--Fairly short, wide snout, dished face, wide between ears.

EARS.--Falling forward over face, long and thin.

JOWL.--Heavy.

NECK.--Fairly long and muscular.

CHEST.--Deep and wide.

SHOULDERS.--Deep and sloping into the back, blades not prominent, but in line with ribs, not too wide on top.

BACK.--Long and level or with a slight arch from head to tail.

RIBS.--Deep and well sprung.

LOINS.--Broad and strong.

SIDES.--Deep.

BELLY AND FLANK.--Full and thick.

QUARTERS.--Long and level or with only very slight droop.

TAIL.--Set high, not coarse.

HAMS.--Very large and well filled to hocks.

LEGS.--Short, straight, and strong.

COLOUR.--White.

SKIN AND COAT.--Smooth; hair straight, fine, and silky and not too much of it.

SIZE.--Large without coarseness.

DISQUALIFICATIONS.--Black spots, black hair, prick ears.

OBJECTIONS.--Blue spots.

* * * * *

[Illustration: From a Painting by Wippell.

A BERKSHIRE SOW.

To face page 32.]

* * * * *

[Illustration: Photo, Sport and General.

LARGE BLACK SOW, "SUDBOURNE SADIE."

Owner, K. M. Clark. 1st Prize and Champion, R. A. Show, Norwich.

To face page 33.]

CHAPTER III

CROSS-BRED PIGS

This term has a varying meaning to different persons. There are those who term a pig a cross-bred unless it be bred from parents of recorded pedigree, or those which possess pedigrees capable of registration. Others claim that a cross-bred is any pig which is bred indiscriminately from boar and sow of no particular type or breeding--in fact common pigs of the country; whilst still others declare that the title of cross-bred can be legitimately applied only to a pig whose parents were of two different pure breeds in contradistinction to a pig sired by a pure bred boar, and from a common sow, or the diverse way.

It is not for us to determine the knotty point, but we may venture the opinion that the two first definitions of a cross-bred are not convincing to us, since in order to produce a cross-bred it is necessary to have a sire, or a dam, or both of defined breeds. Probably the most correct definition of a cross-bred animal is one bred from the mating of sire and dam of two distinct breeds, but the term is now loosely applied to an animal begotten by a sire or from a dam of pedigree breeding, the other parent being of no particular breed.

This system of breeding has become somewhat common owing to the comparatively small outlay required in the purchase of a boar as compared with the purchase of both boar and sows, and also to the belief which is general that a greater improvement in the produce is noticeable when the boar is pure bred and the sows of ordinary or no particular breed, than if the sows are pure bred and the boar a common bred one. In addition to this there is the important point that the pure bred boar should be able to beget at least fifty litters in a year whereas the pure bred sow will not produce more than two litters annually, so that the advantage obtainable from the outlay on one pure bred boar is twenty-five times as great as is possible from the purchase of a pure bred sow.

There is also another advantage to the owner of a boar who has only a limited number of sows, he can allow his neighbours to make use of his boar on payment of a liberal service fee, which combined will partially pay for the prime cost of the boar.

A considerable number of pig breeders are influenced in the purchase of a pure bred boar rather than of a sow by the belief that pure bred sows are neither so prolific nor such good mothers as are common bred sows. This belief was even more common in years gone by than it is at the present time, and it must be candidly confessed that there existed substantial grounds for it. Some fifty years since it became fashionable, particularly amongst those who had suddenly become rich by trade or in other ways, to exhibit live-stock at the agricultural shows. They may have been animated by the laudable desire of endeavouring to assist farmers and stock breeders generally, or a desire to gain a place in the sun may have had some slight influence. As the majority of these exhibitors of stock had no special knowledge of stock, they were compelled to place themselves entirely in the hands of their managers and stockman, who generally received by arrangement a certain percentage of the prize money won by the stock. It was then only natural that they gave far more attention to the show points of the animals in their charge than to the breeding qualities.

The supply of pedigree animals was also very limited at about the period mentioned so that it was much more difficult to avoid too close breeding, nor was there the same care taken in the private record of the pedigrees of the animals bred. These various causes combined led to a loss of vitality amongst the so-called pedigree stock, and this weakening of the constitution showed itself in a reduction in the number of the offspring and in the power of the dam to furnish its young with a full supply of well-balanced milk.

There is little doubt that in the third quarter of the past century a considerable proportion of the pedigree sows were not so prolific as they ought to have been, nor did they produce and rear thoroughly well so many pigs at each litter as the common sow of the country was capable of doing. A more general study of stock breeding has tended to compel attention to the practical apart from the show points of pedigree pigs, but probably the strongest influence has been the formation of the various breed societies, and the registration of the produce including the number, sex, and sire of the pigs. These entries most clearly showed those breeders of pigs who had paid most attention to the utility points of their pigs, especially those particular points in which pedigree pigs were generally believed to be deficient. The succeeding records of sows of the same families afforded the best possible confirmation of the belief which was becoming general that prolificacy like many other qualities was most certainly hereditary. This recorded proof that pure bred animals and especially pigs were not necessarily slow breeders, helped vastly to increase the demand for pedigree animals for crossing purposes in the breeding of commercial stock.

The enormous benefit which has resulted from the use of pedigree sires is most clearly proved in the Irish live stock. The so-called premium bulls and boars are pedigree animals purchased by or with the sanction of the Live Stock Commissioners and placed at the service of the general public at a somewhat reduced fee, the Government paying to the owner an annual premium of some ?5 for each bull, and a certain sum for each boar.

It is alleged that the original improvement in the ordinary pig stock of those parts of Ireland where pig-keeping on a considerable scale is followed, was due to the purchase in England of numbers of Large White boars, as after experiments carried out in Denmark, these boars were found to effect the greatest improvement in the common country pigs and to render them far more suitable for conversion into the kind of bacon which was in most general demand, and of course realised the highest price. For the beginning of the vast improvement in the Irish pig which has followed the importation of these Large White boars, the Irish bacon curers must receive the credit, as they joined together in the purchase of these boars which were distributed in those districts from which the various factories drew their supplies of fat pigs.

A similar plan was adopted by Messrs. Harris of Calne who purchased some hundreds of boars of the Large White breed, and at first lent them on certain conditions to pig breeders, but later on resold the young boars by auction for whatever they would fetch, their object being to secure the use of these boars in order to render the farm pigs more suitable for the purposes of their trade as bacon curers.

There may or may not be any grounds for the belief that the sire has a greater influence in the external form of the joint produce than does the dam, but this belief has also had its influence in determining breeders of cross breds to use the pure bred sire on the ordinary stock of the country, rather than the reverse way. There is no doubt that apart from the improvement in the general quality of the produce of the pure bred sire there results a general uniformity of the young stock, which is a great recommendation when they are placed on the market either as stores, or when fattened for the butcher or bacon curer. This uniformity of type and character in the young stock would be more noticeable still if the buyers of the pure bred sires were to continue their purchases from the same herds, providing that the owners of them were sufficiently careful in avoiding incestuous breeding.

So many people appear to be content with the knowledge that the sire which they are purchasing has a recorded pedigree and is a pure bred sire eligible for entry in the herd book of its breed, but they forget that it is possible in the crossing of two pedigree animals of a similar breed to obtain as great a mixture of blood and points as in the mating of two cross-breds or two come-by-chances. Uniformity in a herd, stud, or flock can only be rendered comparatively certain by the continued use of sires of similar breeding. In making a compound, its character is determined by the proportion of the various ingredients used in its manufacture. So it is in the breeding of stock, those points which are most predominant in proportion in the blood of the sire and dam will, on the average, be represented in an equal degree in the joint produce. This it is which renders so impressive a sire which is descended from closely bred parents. Each of its forbears has handed down a proportion of its own particular characteristics so that the larger the number of animals amongst its forbears which possessed these particular points the greater the certainty of their being possessed by the produce. The meaning of this may be made more clear by pointing out that the result of the mixing together of various mixtures will depend entirely on the proportion of the substances used in the manufacture or compounding of those mixtures. In each animal is embodied the characteristics of its forbears.

There exists generally an opinion that the produce of two parents of distinct breeds, or as it is termed a first cross is commonly superior to a pure bred of either of the two breeds represented by the parents. It is difficult to discover the cause of this, if it be a fact. If one of the parents were deficient in stamina the produce might conceivably be more robust, and it might also occasionally happen that the mixture of the qualities or properties possessed by the parents would result in improvement, as happens when a distinct new breed is originated; but as a rule the good and the bad qualities of the produce from the mating of two animals of diverse breeds are in direct proportion to the qualities possessed by the parents.

The mere mixing of the blood of two animals differently bred cannot increase the good or bad properties, but the combination might possibly result in a blend more suitable for the purpose in hand.

Another claim commonly made for the crossing of animals is that the risk of that delicacy of constitution which they assert is far too common amongst pure bred animals, and is due to close breeding, is hereby avoided. It must be admitted that in times past there was a certain amount of cause for this complaint of want of constitution amongst pedigree animals, but the cause has been considerably if not entirely removed by the more careful recording of the breeding, and by the more drastic screening out of any animals suspected of delicacy of constitution.

The buyers of pure bred animals for crossing purposes have also become more careful in their selection. They have ceased to imagine that because the owner of certain animals most of which he has purchased is successful in winning prizes at the chief agricultural shows, the whole of the animals in his stud, herd, or flock must be of equal excellence or at all events sufficiently good for the production of profitable commercial stock. Action on this mistaken belief has led to much disappointment in the past, since the home bred animals may have been of totally different blood from those which have won prizes, and further they may not be inbred for a sufficiently long time on distinct lines to render them prepotent enough to impress their good qualities on their produce.

Amongst the objections made to cross-breeding is the heavy cost of replacing the breeding stock, as to obtain a first cross, a succession of sires and dams must be purchased. Many persons meet this difficulty by merely buying sires of a breed similar to the first used, but then the produce ceases to be cross-breds and become grades until such time as by the use of a certain number of sires of a similar breed the produce become eligible for entry in the herd book of the sires which have been continuously used. This system of breeding insures a greater uniformity in the produce providing that the sires selected are of similar breeding, type, and character, than even by the system of crossing sire and dam of two pure breeds.

The risk attending too close breeding as in the breeding of pure breds is also avoided provided that the herd from which the sires are bought is sufficiently large to furnish a change of blood, yet of similar breeding.

No one possessing a knowledge of the ordinary farm stock of the country will for one moment deny that there is still vast room for improvement in our live stock, and particularly in our pigs, and it is equally the fact that our Government has not shown a readiness equal to that of some foreign Governments, and even of the authorities in some of our colonies to assist farmers in obtaining the use of improved sires. Take Canada as an instance. For years the Dominion Live Stock Branch has been purchasing and delivering free into districts needing them, male animals for the use of farmers and stock owners free, save stallions, for which a covering fee has to be paid sufficient to cover the insurance of the stallion. The other important condition which relates to all the sires provided by the authorities is that the cost of maintenance shall be paid by the Local Association which has the management of the sire and the arrangement of its services.

Another noticeable point is that all the sires allocated to the various districts are Canadian bred, and so far as is possible are purchased in the province in which they are to be located. The object is undoubtedly to encourage in Canada the breeding of pure bred animals and may thus far be considered satisfactory, but it is acting on an assumption which may not be justified that there exists in the Dominion a sufficiency of stock equal in quality and breeding to those which it may be possible to import.

Within the past three or four years our Board of Agriculture have taken some steps to assist our farmers to improve their stock. The assistance has taken the form of offering premiums of fixed amounts to private persons or associations who hired or purchased approved stallions, bulls, and boars which were placed at fixed fees at the service of the stock of the public. Already great benefit has been derived from the use of the stallions and bulls, and this to a far greater extent than in the pigs, as owing to an unfortunate condition which was attempted to be enforced as to the formation of pig clubs and impracticable conditions the number of boars located in the country has been much smaller than would have been had the conditions at present in force been adopted at the initiation of the scheme.

The boar conditions are now of a similar character to those in force from the first with regard to stallions and bulls. In addition to the supply of male animals at comparatively low fees an attempt has been made to assist in the recording of the milk yield of cows, a matter of the highest importance. If only this could be extended to sows there would soon cease to be cause for the far too common complaint of the owners of sows of certain breeds of pedigree pigs, as to the limited quantity of milk which is provided by the sows for their litters of pigs.

* * * * *

[Illustration: Photo, Francis Davis, Needingworth.

THREE MIDDLE WHITE BREEDING SOWS.

The Property of the Author. Also portion of 17 Sties at Holywell Manor, near St. Ives.

To face page 48.]

* * * * *

[Illustration: Photo, Sport and General.

A MIDDLE WHITE BOAR.

From the Author's Pig Farm.

To face page 49.]

CHAPTER IV

DENTITION AND AGE OF PIGS

Although the majority of pig sellers may claim to be, and may be able to substantiate the claim to be, equally as honest as the majority of others in trade, yet there may be a small minority who are apt to attempt to palm off pigs as being older than they really are. It is most annoying when you are anxious to purchase pigs of say six or seven months old which are quite ready to be quickly fattened, to have pigs of four or five months old which continue to make growth instead of flesh, so that they are not ready for killing until two or three months after they are required for conversion into bacon.

Although the object of the Council of the Smithfield Club is to prevent fraud of a different character, i.e. the exhibition in classes limited to certain ages of pigs of an age greater than that given on the entry form, yet the following table showing the normal state of the dentition of pigs at certain fixed ages will enable purchasers to discover whether or not the seller has attempted to deceive him. It may at once be admitted that there will be a limited number of cases in which the state of dentition of pigs is abnormal, but after examining the teeth of some thousands of pigs during the past sixty years, we have no hesitation in asserting that more than half, at least, of the variations from the normal are allayed dentition. It is claimed that a man of experience is quite able to arrive at the approximate age of a pig by its development and appearance; some few persons may have that instinctive knowledge more or less fully developed, but this examination of the state of dentition is of the greatest possible assistance in arriving at the actual age of the pig, particularly desirable as it is in case of a difference of opinion between buyer and seller.

The following are the conditions of the state of dentition to which all pigs have to conform ere they are allowed to compete for the prizes offered by the Smithfield Club at their annual shows:--

"Pigs having their corner permanent incisors cut will be considered as exceeding six months.

"Pigs having their permanent tusks more than half up will be considered as exceeding nine months.

"Pigs having their central permanent incisors up, and any of the first three permanent molars cut, will be considered as exceeding twelve months.

"Pigs having their lateral temporary incisors shed, and the permanents appearing will be considered as exceeding fifteen months.

"Pigs having their lateral permanent incisors fully up will be considered as exceeding eighteen months."

As the majority of the pigs bought of dealers by amateurs are young pigs it may be advisable to state that a pig of the age of eight weeks old should have its two central incisors fully grown. A pig three months old should have all four temporary incisors cut, the two outside ones being more than half as long as the two central incisors.

As the first set of the teeth of a pig like that of a child are merely temporary, and as these give place at fairly definite ages of the owner to permanent ones, it may be well to endeavour to describe as clearly as possible the position and appearance of the temporaries as compared with the permanents. The pig is one of the few animals which is possessed of teeth at its birth; these number eight, two on each side of the upper and lower jaw. It has been suggested that these early teeth are provided to assist the pigling to grasp firmly the sow's teat when in the act of sucking. These eight teeth vary somewhat in length; those pigs which are carried by the sow beyond the usual period of sixteen weeks frequently have longer and even sharper teeth than those of pigs which are born at the usual time. These longer teeth are also sometimes of a dark colour. This is doubtless the origin of the remark commonly made by old-fashioned pigmen that "pigs born with black teeth never do well." This might have been so prior to the discovery that the breaking off the sharp teeth of the newly born pigs frequently saved trouble, and often the life of the little pigs. Pigs whose teeth are discoloured at birth are usually more robust rather than the reverse, since the sow carrying them beyond the allotted time is invariably in a vigorous state of health, and her pigs consequently more fully developed.

When the pig is about a month old, the two central incisors are cut in each jaw, these are two of the four front teeth in each jaw of the pig at a subsequent age. Two temporary molars are also cut on each side of the jaw above and below, with the first temporary molar in each place ready to come through the gum.

At two months the temporary central incisors are fully developed, and the two lateral temporary incisors can be seen in the gums, if they are not already through. All three temporary molars are now about level.

When the pig is about three months old its temporary teeth are all in position, the temporary lateral incisors are through, and nearly as long as the temporary central incisors. Owing to the lengthening of the jaws the two temporary corner teeth which were present at birth will have become further apart. When the pig is about five months, the fourth molar in either jaw shows itself in the gums, then at six months the wolf teeth show between the tusks and the premolars, and the fourth molar is nearly level with the first premolar. The corner incisors and the tusks usually disappear, and are replaced by permanents when the pig is nine months old. The second permanent molar also shows itself. At twelve months the two central temporary incisors give place to the permanents; these last are more square in form than the temporaries, and are thus easily distinguished. The three temporary molars will also be ready for displacement by three permanents. These last will be level with the other permanent molars when the pig is fifteen months. The two lateral incisors will also have given place to permanents. At eighteen months the third permanent molars will be coming through, and at the age of twenty months the pig's teeth are fully developed.

CHAPTER V

SELECTION OF THE BOAR

The hackneyed saying "The sire is half the herd" appears to have a different meaning to varying persons. To some it conveys the idea that the selection of the sire is of far more importance than the selection of the dam because the influence of the sire is so much more powerful than that of the dam on at least the external form and character of the produce. The late Mr. James Howard, who took a particularly keen interest in the breeding of pigs, used to declare that the appearance and form of the young pigs far more generally followed those of the sire than of the dam; whilst the influence of the latter was more shown in the character and constitution of their joint produce; or in other words, that the boar stamped his character to a greater extent on the external points of the young, whilst the sow more strongly influenced the internal parts of the youngsters. It is quite possible that this idea has gained ground to a large extent from the fact that the use of a pure bred sire on ordinary or grade females has been very much more common than the crossing of pure bred females by the ordinary or non-pedigree sire; as also from the far greater numbers of young which each pure bred sire would improve, than would be improved by each pure bred female which might be crossed.

If only for this reason alone, we would always recommend buyers who are desirous of grading up and improving their farm stock to attempt to do this by the purchase or use of the pure bred or improved sire. The original outlay is infinitely less, whilst the immediate results are comparatively longer.

It is scarcely desirable to go further into the question as to the comparative influence on the young of the sire and the dam since our actual knowledge of the subject is by no means large. Indeed, it is at the least doubtful, if by the closest observation any definite opinion on the subject is possible, so great is the difference which varying parents have on the chief characteristics of their joint progeny, and even in the separate specimens which they have procreated. Of course, it is quite possible to breed animals especially well developed or endowed with certain qualities, providing that the parents have been for generations selected because of their possession in a marked degree of those particular qualities sought. It is in this power of prepotency that one of the chief benefits from the use of a pure bred sire or dam arises. By the term pure bred is not meant merely that the names of a certain number of the forbears of the animal shall have been recorded in the register of the breed, but that the animal shall for a certain number of generations have been bred on similar lines so that it shall possess a considerable amount of concentrated blood. This is a point to which sufficient care is not generally given by purchasers of so-called pedigree sires to be used on the ordinary bred or graded stock. The far too common practice is to purchase each boar required from a totally different herd, or from one of quite dissimilar breeding, with the result that there is not the slightest uniformity in the appearance or character of the herd, or of the mature animals when ready for market.

It is far too frequently forgotten that the chief value of a record of the pedigree is that by it one can trace the breeding of the animal's progenitors, and thus one is enabled to form some opinion of the probable produce-- providing it is possible to learn the chief characteristics of the progenitors. Failing this, the only course open is to note the names of the breeders of the more recent parents, as from this a certain amount of information as to the probable qualities of the parents may be obtained or surmised.

Another point on which at least a diversity of opinion exists, is the wisdom of giving so much consideration to the fact that the herd from which the sire is purchased shall have been recently successful in the show yard, or in extreme cases, that the sire itself shall have been a prize winner. It is urged that the mere fact that a sire has succeeded in winning one or more prizes is a proof that it possesses in a marked degree those qualities which are most highly prized. This may be conceded, yet there is no certainty that the mating of this winning sire even with dams that have also been prize winners shall result in the production of young the equal of the parents, since the winners at the various shows may be of dissimilar types and breeding.

But the case would be quite different if the winning sire and dam came from the same old established herd in which the animals had been bred for generations on similar lines. It is this concentration of certain qualities in generation after generation which renders the pedigree animal so intensely prepotent, particularly when mated with animals of an ordinary character or not possessing concentrated breeding. Indeed, it may be safely assumed that the power of a parent to impress its own individuality and qualities on its produce, depends to a very large extent, if not entirely, on the comparative hereditary extent of those qualities in comparison with the other qualities possessed by itself, or by the animal with which it may have been mated.

In other words, it is contended that the sire or the dam has not the power to impress certain of its characteristics on its young, merely because of its sex, but that this power depends on the proportion in the sire or dam of the blood of progenitors who possessed in a marked degree certain qualities.

It is with the breeding of animals as with the manufacture of a compound article. The character and quality of that compound will vary according to the proportion of the various ingredients used in its manufacture. It is to this law or fact that the marked impressiveness of certain strains of blood is attributable.

Again, the marked and long continued success of the blood of the animals bred by a few of our most successful breeders of live stock is in the main due to the fact that the owners set up a standard and persistently selected and bred together only animals possessing to a greater or lesser extent the particular qualities which together comprised that standard. There is not the slightest doubt that in carrying out their system they were often compelled to mate animals related in blood the one to the other, but in this there is little risk providing that all those animals which show the slightest symptom of delicacy of constitution are persistently draughted out.

It will be inferred from the above remarks that we hold to the belief that the breeding of the boar should receive attention as well as the following points in its form and character.

One of the most important of these points is good temper. This is a quality not usually attributed to the pig in its wild state, and consequently not natural to the domesticated pig, yet on the possession of it depends to a very great extent the thrift and well doing of the produce of the boar. The produce of an irritable boar are almost certain to inherit this quality which is fatal to profitable fatting. In sows this weakness is still more unfortunate, as a bad tempered sow is almost invariably an indifferent mother. The rigid avoidance of this failing of bad temper in a boar is advisable not only because this quality is almost invariably hereditary, but a savage boar is a continual source of danger to man and beast. It may be said that little trouble is likely if the boar is kept in confinement, but there are times, such as when sows are placed with him, when a certain amount of liberty must be given to him, and it is generally on such occasions of excitement when the bad temper is the most in evidence. The mere fact that irritability and nervousness are natural to the pig should make us the more careful to avoid any increase in the failing by using a boar which is the least inclined to be bad tempered.

Many persons hold that in the selection of a boar one of the principal points is size. They contend that size, in pigs especially, is imperative if a profitable return is to be made. This view may have arisen to a greater or lesser extent from the want of method and observation which is characteristic of so many stock owners. The one point which to them is of the greatest importance is the selling price of the fat or store animal sold being fully up to the average. Little or no thought is given to the value of the food eaten by each animal. If it had been, very frequently it would have been found that the smaller animal of a lot had actually given the best return for the food it had consumed. It is not the size alone of the animal which determines its value as the producer of meat, but more than anything it is the feeding qualities of the animal fattened. In addition to this there never was a time when the consumer more strongly demanded small joints of meat, and these of the best quality and with as little bone as possible.

Apart from this a very large boar is a mistake as it is invariably awkward when serving--it can be used only for large and strong sows, and its average period of usefulness is decidedly shorter than that of a medium sized and compact boar. A large boar generally possesses an undue proportion of bone, its shoulders are heavy, and its ankles round, and feet large and spreading. Now these are all objections. The bone of a boar should be solid, not porous;

the ankles compact and the feet small, and the pasterns short. The head should be wide so that the brain can be well developed, the head inclined to be short rather than long, since an animal with an extremely long head is certain to be deficient in natural flesh.

On the question of the size and hang of the ears a variety of opinion exists; pigs with long ears, and pigs with short ears are found possessing good carcases. It is the quality of the ear rather than its size and hang which seems to indicate the character most. A pig with a thin and firm ear is usually of fine quality, whereas a pig which has a thick, coarse ear is generally coarse in bone, skin, flesh, and hair.

The neck of the boar should be muscular as indicating constitution and natural vigour; the shoulders fine and obliquely laid, the ribs well sprung, the loin wide, the quarters long and square, not drooping, the hams full and extending quite down to the hocks, and without any of that loose skin which is far too common amongst the largest of our breeds of pigs, and which is a sure sign of coarseness. The flank should be thick and well let down, as this indicates constitution and lean meat, the legs should be fairly short and set well apart so that the heart, lungs, and other organs have plenty of room to perform their share of the work of the pig. The skin should be fine and the hair straight and silky, as well as plenty of it. Sparsity of hair is generally an indication of shortage of lean meat, whilst curliness and coarseness of hair are far too frequently associated with excessive fat and coarseness of meat.

With regard to the reproductive parts of the boar there are one or two points which should receive special attention. A boar with excessively small testicles should be avoided, as such a one is often barren. Again, a boar with one testicle of normal size and the other smaller, ordinarily suffers from the same disqualification to a lesser extent. A ruptured boar should not on any account be used, as this weakness is strongly hereditary. The weakness may not possibly show itself in the first generation, but it is certain to appear sooner or later. Not only is it a sure index of weakness of constitution, but pigs so affected occasionally die suddenly, whilst there is always a certain amount of risk from the operation of castration.

Occasionally one or more of the boar pigs of a litter will be found to be malformed, in that only one of the testicles is apparent. Generally speaking, the other is found when the pig is killed to be attached to the inside of the pig, and thus is unable to descend into the scrotum or purse, so that the act of castration is only partially performed. A boar pig with only one testicle down is commonly termed a rig. The removal of one of the testicles does not deprive the rig of reproducing its species, and it is thus a source of continual trouble when herded with a lot of sow pigs now that the general custom is to allow the female pigs of a litter to remain unspayed. It is, therefore, necessary to fatten a rig either alone, or with male pigs which have been operated upon. In addition to this extra trouble, the flesh of a rig pig if it be kept fattening after it is some five or six months old is almost certain to be strong in flavour, like unto that of a boar. It is, therefore, advisable to fatten a rig quite early in life and convert it into a porket or porker carcase of pork.

It may appear strange to some readers to specially mention the teats of the boar, but it is equally as necessary to avoid boars having small teats, teats unevenly placed, and commencing any distance from the fore legs, and blind teats, as it is in the case of the sow, since any weaknesses of the kind are equally as hereditary from the boar as from the sow.

CHAPTER VI

SELECTION OF THE SOW

It is impossible to agree with the view held by so many persons that the necessity for the same care is non-existent in the selection of a sow as in the choice of a boar. We hold that the desirability for studying the forbears, especially the dam, of a young sow intended for breeding purposes is fully as great as when selecting the young boar, since many, if not most, of the qualities which we desire the brood sow to possess are strongly hereditary. Take, for instance, the question of gentleness or a quiet disposition, it follows from dam to produce with a regularity equal to that of bad temper, and the latter is wellnigh a certainty. Again, whoever found that the female produce of a sow deficient in the maternal instincts proved, if saved for breeding purposes, to be a really good mother? As a rule the daughters of a sow which gives but a small quantity of milk, and that of an inferior quality, are also cursed with the same grievous failings, but this does not appear to be universally the case, since the milking qualities of the dam descend through her sons, so that if the female progenitors of the boar have been good milkers it is probable that the boar's daughters may be able to rear their pigs successfully, even if their dam had not been in the habit of suckling her pigs well.

Still, it is quite safe to assert that with this one exception we may fairly anticipate that the good qualities which we seek in a sow are far more likely to be found in the sow pigs of a sow herself the possessor, than from one which does not possess them. We are inclined to the belief that the alleged failure of some pedigree yelts to make good brood sows is in the main due to the continued selection for breeding purposes of those pure-bred yelts which show early maturing and flesh-forming qualities, rather than that motherly look which is almost invariably to be found in a sow which is prolific, a free milker, and a really good mother. There is a marked difference in the formation of a milk-giving and a fat-producing sow--the latter is generally somewhat heavy in the shoulders, has a muscular or fat neck, is rather short in the head and heavy in the jowl, and is altogether more compactly built, whereas a good brood sow has rather a long face, is wide between the eyes, has a light muscular neck, is fine in the shoulders, possesses long and square quarters and appears to be heavier in the hind than in the forequarters. She is somewhat more loosely built and often shows less of quality. Thickness of flank and length of side are desirable, the first as indicating substance and flesh, whilst the second gives plenty of room for her pigs to suck. The bone should be of good quality; the same remarks apply to the skin and hair.

About half a century since there existed a fancy, which almost amounted to a craze for sows of small size; they could not be too neat, and showing too much so-called feminine character. The almost certain result of selecting the neatest of the female pigs followed, the fat pigs sent to market were light in weight, deficient in lean meat and rightly named "animated bladders of lard." Within about the same distance of time it was the common practice of exhibitors of pigs at the Smithfield Club's Shows to provide pillows in the form of round pieces of wood on which the fat pigs rested their heads so that these were raised in order to prevent the pigs becoming suffocated. In addition, the pigs were fed on forcing foods until they were at least one and a half year old and allowed to take, or were given little exercise, with the result that the pork consisted mainly of soft fat or lard. To such an extent had this craze for neatness been followed that the bacon curers and consumers of pork wellnigh ceased to purchase or consume pork.

At the present time we are afraid that the tendency is in the opposite direction, and mere size is receiving far too much attention. At some of our agricultural shows the judges select for honour great unwieldly sows which could not possibly perform with any amount of success those maternal duties which a brood sow is supposed to be kept solely to perform. An extremely large sow is very frequently a poor milker, the quantity of milk she gives is not large, nor does she continue to give even this reduced supply for a period long enough to allow her young to grow strong enough to make a good start in life on their own account.

Another great objection to a sow of extreme size is that her produce almost invariably take after her to such an extent that it is difficult, if not wellnigh impossible, to make them fat until they are from nine to twelve months old, and by that time they are too large and heavy for the general demand which is at the present time, and likely to become still more so in the future, for small joints of meat which carry a large proportion of lean and a limited quantity of bone. The most successful manufacturer is he who most nearly supplies the consumer with that which he requires or fancies. We are not moved by the contention of breeders of pedigree pigs that the most valuable pig is the one which possesses in the greatest degree those special points which are characteristic of the breed, as, for instance, size in the pigs of the Large White, the Large Black, and the Lincolnshire Curly Coated pigs, therefore the biggest pigs should be held in the highest esteem. In our opinion the best, as it is in the long run the most profitable, is the pig which furnishes to the greatest extent exactly the kind of meat in the most general demand.

In addition to these objections to an extremely large and ungainly sow is the fact that such an one is invariably clumsy in the breeding pen, she is almost certain to lay on some of her pigs. It is even alleged that her period of usefulness as a breeder is shorter than that of a sow of ordinary size.

* * * * *

[Illustration: Photo, Sport and General.

TAMWORTH BOAR: BISHOP OF WEBTON. Owner, C. L. Coxon. 1st and

Champion, Royal Show.

To face page 64]

* * * * *

[Illustration: Photo, G. H. Parsons, Rostrevor.

GLOUCESTER OLD SPOT SOW. From the herd of Lord Sherbourne.

To face page 65.]

CHAPTER VII

THE SOW'S UDDER

One of the most important points in connection with the reproduction of the species of our various domestic animals is the provision of a full supply of milk for the young in the early portion of their existence. Nature herself has set us a good example in a duplicated source of milk supply even amongst animals which usually produce only one animal at a birth. If this duplication be necessary under such conditions, it must be imperative on us to select those sow pigs which are intended for breeding pigs which possess a well-formed udder, having a full supply of teats, and these of good shape and properly placed on the belly of the sow. Not only is this necessary to ensure the rearing of a fairly numerous litter of pigs in a satisfactory manner, but it is held that the number of teats possessed by a sow indicates to a remarkable extent the probable degree of prolificacy of the sow. One can readily understand that nature would not be likely to endow a sow with the power to produce a larger number of young at each birth than she would be able to rear. Of course it may be said that the sow of the present day is not as nature first made her, in that, by selection and by feeding, the number of pigs produced at each birth is now so much larger than the litters of the wild sows, which have some seven or eight teats and farrow at each litter a similar number of pigs. On the other hand, neither the number of teats nor of the young is fixed either in the domesticated sow, or the sow in a wild state, so that by continued selection we are able to permanently increase, within limits, the production of larger litters and the increased supply of milk for their sustenance when young.

The provision of a suitable udder is even of more importance with the domesticated than with the sow running wild, since the latter produces each year one litter only, and that in the season of the year when the young are less dependent on their dams; whereas the domesticated sow is expected to rear at least two litters per year, and frequently owing to want of care on the part of the owner the young pigs are farrowed at the most unfavourable time of the year.

Apart from the provision of a certain number of teats there is another point to be considered, the power of the sow to produce milk enough to satisfy the given number of pigs. This of course varies with each family or tribe of pigs, and even with the various members in it, so that to obtain the best results selection must be made of the produce of those sows which give the largest quantity of the most nutritious milk. There exists amongst pig keepers a difference of opinion as to the number of pigs each sow should be allowed to rear, probably the average of this number would be ten for a mature sow, and seven or eight for a first litter. If so, the selected sow pig should possess twelve teats, as frequently one of these may not give a full supply of milk from either natural or accidental causes. The teats should be regular in size and form. It is not uncommon to find one or more of the teats of a sow much smaller in size than the others. These smaller teats will produce a reduced quantity of milk, so that the pigling which is unfortunate enough to decide on making this small teat its very own--and each pigling is confined by the others to its own special teat--is certain to be less well developed than its brothers and sisters, even if it should succeed in surviving.

The necessity for the teats being placed equidistant the one from the other arises from the desirability of each pig having room to suck comfortably. Should two of the teats be closely placed the two pigs will probably fight, when not only will sores be caused on the cheeks of the pigs, but the milk in the teats not properly drawn will gradually cease to flow.

Another point of great importance is that the teats should commence as near as possible to the fore legs--this for two reasons: it gives more room for the pigs to suck as they grow larger; the other and more important one is that the teats most forward on the udder of the sow produce the larger quantity of milk, or milk of a better quality. It will be almost invariably found that the pigs sucking the foremost teats thrive the best.

It is advisable to avoid the selection of a female pig for breeding purposes from a sow which has large and coarse teats, as these invariably accompany coarseness of skin, bone, hair, and flesh, this in turn affects the sale value of the carcase. It will also be found that those sows with a neat and compact udder, with fine teats, will give more milk and a better quality than sows possessing a coarse and flabby udder.

Another weakness to be avoided when selecting a sow pig for breeding purposes is that which is commonly termed a blind teat, since it is undoubtedly hereditary in addition to being useless for the purpose of rearing pigs. The normal teat projects boldly from the udder, whereas the blind teat is almost flat or on a level with the udder. In appearance it resembles a ring of skin with a depressed nipple in its centre. At the time of parturition the blind teat contains milk to the same extent as do the other teats, but it promptly dries up since it is impossible for the little pigs to extract the milk from it since the nipple recedes as soon as the pigling attempts to clasp it with its lips and tongue, instead of becoming more extended so that the little pig can suck the milk from it.

An ideal udder can be briefly described as one possessing at least twelve fully developed teats, the more the better--these should commence from a point as near the fore legs of the sow as possible, and be placed as nearly as possible an equal distance the one from the other.

Some persons hold that large teats and much loose skin are sure indications that the sow has proved to be a good milker. This is a mistaken view; it is with sows as with cows, the most prolific milkers are those with well formed and soft udders which almost disappear when the lactation period has passed.

CHAPTER VIII

MATING THE YOUNG SOW

As in most other details in the management of our domesticated animals there exists a variety of opinion as to the age at which the young sow, or, as it is commonly locally termed, a gilt, yelt, yilt, hilt, elt, etc., should be mated with the boar. Perhaps the most important point to consider is the time of the year when the anticipated pigs should arrive. If possible the period between the middle of the month of September and the middle of December should be avoided. The long nights and the short and dull days generally experienced during this period are most unsuitable for young pigs. Many litters of pigs farrowed in October are not any larger nor nearly as thrifty in the month of March as those farrowed during the early portion of January in the following year, and very frequently the loss amongst the October and November farrowed pigs from lameness, or, as it is commonly termed, cramp, is very large. An attempt should be made so that the two litters which should be bred each year from the sow arrive so that they are weaned during the longer and brighter days of the year; thus a sow which farrows in the beginning of February may be expected to farrow again in July. The weather is sometimes rather cold for little pigs early in the year, but it is surprising how little they are affected by it providing the sow furnishes a good supply of milk and the bedding is dry and plentiful. The pigs farrowed in the months of January and February are generally the most profitable, as they will be ready for the consumption of the separated milk, butter milk, whey, etc., which is generally abundant in the month of May in districts where dairying and cheese making are followed. These pigs are also ready for turning out to grass in April or May, or as soon as the weather is suitable, and the grass has grown sufficiently. These young pigs will grow and thrive splendidly providing that some additional food is fed to them and shelter provided.

This natural system of pig raising is of great benefit to those pigs which are intended for breeding purposes and was consistently followed by the writer from the year 1863. It was by no means a new plan even at that period, although strange claims have recently been made that the system is a novel one and originated in the fertile brain of one or two enthusiasts who have gone the whole hog in pig breeding. In the middle of the last century it was quite a common practice in parts of the counties of Cambridge, Essex, and Suffolk to graze the seeds which comprised clovers, trefoils, etc., with pigs which received in addition extra food, such as peas or beans in accordance with the amount of vegetable food obtainable or the purpose for which the pigs were required; those intended for pork receiving the larger supply.

Although it may not be possible to allow the young boar pigs to have their liberty after they become five months old, yet the sow pigs will grow and develop far better in the field if properly fed than they will in an inclosed sty; further, the young pigs which they produce will be much more lusty than if the sows had been kept in close confinement.

Although the sow pig will generally come in heat when she is about six months old, it is advisable that she should not be mated until she is some eight months old, so that her first litter of pigs is not farrowed until she is about a year old, when she should be quite strong enough to rear a fair litter of pigs and also to grow and develop into a fully natured specimen of its breed.

In some districts where the breeding pigs are generally kept in confinement and high keeping is followed the sow pigs are mated with the boar at an earlier age, but the system has its disadvantages which more than outweigh the saving of the extra few weeks of the keep of the yelt ere she is put to the boar. This early mating is especially harmful if the number of the pigs in the first litter should be large. So few pig keepers have the hardihood to knock a certain portion of the too numerous litter on the head, and so reduce the number to say seven or eight, which most young sows should be able to rear fairly well and without any undue drain on the sow's system--but the whole of the large litter are left on the sow, which becomes very much reduced in condition, and checked in growth, whilst the too large litter of pigs are badly reared and frequently become a source of trouble and annoyance to the owner.

On the other hand, there are many practical pig keepers who make it a rule to delay the mating of their young sows beyond the eight months' age. They contend that a sow pig at eight months is not sufficiently matured to bear the strain of producing a litter of pigs when she is about one year old, and then to furnish the pigs with a sufficiency of milk to give them a good start in life. The plan which they adopt is to mate the sow when she is about a year old so that she is some sixteen months old before her family troubles commence.

Another very curious reason has been recently made public by an enthusiastic novice for delaying the mating of the yelt until she is at least a year old. It is the following, that it is quite possible to ensure that the produce of young sows which have reached the age of sixteen or seventeen months ere they farrow their first litter shall possess the desired characteristics of the breed, whereas this is by no means certain if the young pigs arrive before the sow has reached that age or is about a year old. Unfortunately, we have see no attempt made to account for this alleged curious variation in the qualities inherited from a parent of about one year old and the parent which had arrived at the more matured age of about sixteen months, so that it is impossible to discover a solution of the strange problem.

Therefore, we should be unable to admit the correctness of the assertion even though it was not directly in opposition to our belief which is founded on experience of a most extensive character extending over some sixty years. To aver that the power of a young sow to impress its hereditary characteristics on its young are only fully developed by deferring the arrival of the young pigs for four or five months, or until the sow is sixteen instead of twelve months old when she farrows must surely cause surprise, if not, disbelief. Perhaps the object of the propounder of the theory was to create a discussion--it could not have been to bring his name prominently before the public.

Another advantage in mating the young sow so that her first litter appears when she is about a year old, her daughters will in turn farrow during the most suitable months of the year, providing of course that she herself had been farrowed in early spring or about the month of July.

When the sow pig which is intended to be kept for breeding has been farrowed in some other part of the year, it is advisable to defer beyond the eight months the mating of her so that she farrows at the best times, or perhaps even better than that, if the pigs are not intended for breeding purposes, would it be to have the sow mated when she is about eight months old, and then allow the pigs to remain on the sow a few weeks beyond the usual period of eight weeks so that the pigs are taken off the sow three or four days before it is desired to have her again mated with the boar. The risk of the sow returning to the boar will be minimised, as a sow which has been baulked is sometimes difficult to settle. In addition, the sow will be stronger and more vigorous and likely to produce a strong litter of pigs, whilst the piglings will scarcely miss their mother's milk when they are weaned from her.

Those pig breeders who are in favour of withholding the boar from the young sow until she is about a year old aver that early mating results in the sow becoming worn out and useless for breeding at a much younger age than if she be not mated until she is well matured. This is not in accordance with the writer's experience, as many of his sows which farrowed their first litter when they were about a year old continued to breed regularly until they were six or seven years old--indeed, one Middle White, Holywell Victoria Countess farrowed her last litter when she was in her eleventh year. This sow also disproved the confident assertion that the showing of sows renders them comparatively useless for breeding purposes, since she not only continued to rear her pigs well, but she produced a number of most successful prize winning boars and sows, and also won many prizes herself from the age of five months to five years.

The principal cause of premature old age amongst sows is not due to their being first mated when they are eight months old, but to the want of care in the management and feeding of the sow during her pregnancy and whilst she is suckling her litter of pigs. To a sow with a good constitution the act of breeding and rearing a family of pigs is only the most important act of nature which cannot be harmful to her, providing that she received that amount of proper food and attention which nature required.

There may be very occasional instances of harm being done to the breeding sow by over feeding, or rather by injudicious feeding, but in comparison there are hundreds of instances where under feeding and neglect are the cause of trouble and loss.

CHAPTER IX

THE FARROWING SOW

The pregnant sow usually carries her pigs about sixteen weeks. The variations are neither great nor numerous, when they do occur it is usually with sows with their first litters or aged sows which sometimes farrow ere the full time has expired, or with robust sows in good condition which occasionally carry their young beyond the one hundred and twelve days which may be taken as the average period.

We assume that each owner of a breeding sow keeps a record of the date of service of the sow in order that the necessary preparation of the sty, etc., can be made in readiness for the arrival of the expected litter. Even when this wise precaution is neglected nature gives a sufficient warning to the observant owner. Apart from the increasing size of the body, the udder gradually becomes more prominent, and each pap becomes more defined, the vulva becomes enlarged and the muscles on either side of the tail fall away and lose their tenseness, whilst in the vast majority of cases milk appears in the udder some twelve hours before the arrival of the pigs. The teats shine and become more prominent, the presence of milk is easily ascertained by gently pressing the teat between the finger and thumb.

Another certain indication of the early arrival of a family is the act of the sow carrying straw about in her mouth wherewith to make her farrowing bed.

One of the chief causes of trouble with the farrowing sow arises from the sow not having been allowed to take sufficient exercise. Of course, the best of all systems is to allow the sow its complete freedom at all seasons of the year save when she is within about a fortnight of her time, and when she is rearing a litter of pigs. Even if there be no grass field or paddock in which she is able to pick up a good portion of her living, or a roadside where she can get a few blades of grass, an open yard is infinitely better than the confinement of a sty, as apart from the reduction in the cost of keep, the sow will produce stronger pigs, and have a decidedly easier time of farrowing.

In the case of a sow showing a disinclination to take a sufficiency of exercise either owing to laziness, to high condition, or heaviness of body, it is advisable to exercise her by walking her about quietly for a short time each morning and evening before and after the heat of the day has become excessive, or has passed off.

There is a difference of opinion amongst pig breeders as to the desirability or the reverse of having someone in attendance on the sow during the time she is farrowing. Those who object to this procedure do so on the ground that the presence of a man simply tends to irritate the sow, and to frequently cause her to become restless, with the result that the little pigs are trodden upon or become laid upon and killed. This view is generally held by those who are not particularly fond of animals, as evidenced by that occasional intercourse between pig and owner which consists of rubbing the head of the pig, or scratching its side, when in response to the pleasant sensation it rolls over on to its side like Oliver and asks for more. The trouble if any, which arises from the owner or pig man being in the sty with the farrowing sow almost invariably arises from the absence of a sympathetic feeling between the two. Sows, and indeed wellnigh all animals, pine for sympathy and company, and no animal more so than a pig. Still there are very occasional instances where the young sow becomes very excitable as she commences to farrow even when she and her owner or attendant have previously been on the best of terms. But there the cause is not the presence of a human being, but the arrival of one of her own little pigs. So long as the pigling remains quiet there is peace, but as soon as the youngster endeavours to get to the teat and especially if in the endeavour it utters a cry or a squeak, the young sow will jump up from her nest and endeavour to seize the youngster in her mouth, when unless prevented the sow quickly squeezes all life out of the pig; and in some cases when the pressure has been so severe as to break the skin of the piglet, and the sow tastes blood, she will proceed to eat the dead pig. When affairs have arrived at this sad state, the chances of the remaining pigs having a pleasant reception into the world are comparatively slight.

When there exists a good understanding between the sow and attendant, as there invariably is when the latter is not rough and unkind, as only bad tempered men can be, the trouble in a case such as just referred to is greatly reduced, as the attendant runs no risk in entering the sty and in removing the little pigs as they arrive, and placing them in a hamper or box partially filled with straw until such time as the sow has completed her farrowing, when the pigs can be placed against the sow's udder, and providing they do not bite her, all will settle down in peace and comfort.

In order to avoid the slightest risk of trouble it is advisable when the pigs are apart from the sow to break off with a pair of pliers the four little teeth with which the pigs are endowed prior to their birth. Care being taken to remove the pigs beyond the hearing of the sow each little pig in turn is tucked under the left arm, the mouth is opened by the left hand, and the teeth pressed hard with the pliers, or even a slight turn of the wrist given, when necessary, and the brittle teeth are crushed.

As soon as the placenta or afterbirth is ejected this should be removed. A little slop food should be fed to the sow, and whilst she is eating it, the wettest part of the bedding should be replaced by a little short and dry straw just enough to render the nest comfortable for the little pigs. The nest should be disturbed as little as possible, as should the whole of it be removed and fresh straw given, the sow will probably spend a considerable time in remaking the nest, and in the meantime the little pigs will be in danger from a chill, or in being mixed up in the straw and being laid upon. The risk from delay in the sow laying down and suckling her pigs is much greater in cold weather, as not only will they become chilled, but they will persist in crowding round the sow and so run the greater risk of being trodden upon, or rolled up in the bedding in the effort of the sow to remake her comfortable nest.

Some persons strongly recommend the giving of a strong dose of medicine to the sow after she has farrowed. In ordinary cases this is not necessary, the farrowing of a litter of pigs is a simple and natural operation. In those occasional instances where manual assistance has to be given to the sow owing to the unusual size of the pig, or wrong presentation, or even of a pig which has been dead for a day or two and has begun to decompose and consequently to increase in bulk, it is advisable to give medicine to the sow, since there is every probability of some amount of inflammation due to the insertion of the hand. As the sow's bowels are likely to be somewhat constipated it is always advisable to exercise her for a few minutes during the morning after she has farrowed. In most cases the exercise will at once cause her to relieve her bowels and her bladder, when she can be returned to her sty.

We found sharps, or the finest portion of miller's offals (which usually go by varying names in different parts of the country), the most suitable food for newly farrowed sows, and until the pigs were at least four weeks old. Some persons recommend that a portion of the food should consist of bran, this on two grounds--the first that its use tends to prevent constipation, and secondly on account of the food analysis which it gives. Our experience has been that when sharps are fed to the sow no trouble should arise from constipation, whilst as to the nutriment which bran contains the claim may be good, but the pig is unable to extract it; so large a proportion of the bran passes through the pig in an undigested condition. As a rule the pig, unlike the horse, cow, or sheep, does not masticate its food, nor does it, like the two last named, chew its cud, but it usually bolts its food, and thus casts a greater labour on its digestive organs which have neither the time nor power to extract the whole of the nutriment from the bran. In addition to this, bran tends to too great looseness of the bowels, which in the case of young pigs tends to become diarrh[oe]a.

There is a tendency on the part of some pigmen who are over anxious to succeed to feed the sow too large a quantity of food during the first ten days or so after she has farrowed. During this period the demand on the sow is really not much greater than it was during the last two or three weeks of carrying the pigs. As the pigs grow older an increased supply of food is necessary, but for a week or two after the arrival of the pigs twice feeding of the sow should suffice unless she is very low in condition, or a very large litter of pigs is left on her. In such cases it may be advisable to feed her three times per day just as much as she will promptly clear up. It is a great mistake to give so much food at one time that a portion is left over in the trough, particularly is this so when the pigs are over three weeks old, as nothing so quickly upsets the stomachs of the youngsters as sour food. If in ordinary cases feeding the sow three times daily is persisted in, the same quantity of food given should simply be distributed over the three feedings, as an excess of food is only less a mistake than underfeeding.

Three of the most common troubles with young pigs are diarrh[oe]a, fits, and loss of the tail. There is a belief amongst many old pigmen that every litter of pigs is bound to have at least one attack of diarrh[oe]a ere it is weaned. They look upon it as a fatality which is certain to eventuate, no matter what steps may be taken. Of course, this is folly. The liability of little pigs to an attack of what is commonly called scouring Is great as the causes are several, amongst them the greatest is perhaps a chill which may arise from draughts owing to faulty construction of the sty doors or ventilators neglected; insufficient, unsuitable, or damp bedding; neglect of proper sanitation, or the frequent cleaning out of the sty; and most frequently of all from injudicious feeding of the sow. In fact, anything which affects the health of the young pig to any great extent appears to result in indigestion, which causes constipation, and this in turn nature endeavours to remove by a special effort which softens the f鎓 es somewhat. On removal of the cause of the constipation, the bowels perform their duty normally, but if this be not removed the result is diarrh[oe]a, which again if continued for any length of time often becomes dysentery, when the fever is acute; the pigling neglects its mother's teat, and instead sucks up any moisture however foul which it can find in the sty. This is almost invariably a precursor of death.

From the above it will be gathered that prevention is better than cure. In case of an attack, the first thing is to discover the cause, and the second is to remove it, when, generally speaking, the trouble ceases. As a help to this end, depriving the sow of one meal is recommended. Coal, cinders, or even earth will be readily eaten by the young pigs and prove of benefit. Medicine is not often required if the steps recommended are promptly taken.

In our earlier days hog's madder was the common medicine used with pigs for most ailments, but of late years sulphur appears to have taken its place. It is less violent than castor oil, which is apt to cause constipation of the bowels after its first effect has passed off.

The soreness of the tails, which if not attended to generally results in the pig becoming bob-tailed, appears most generally in damp and cold weather, and is the result of impaired circulation of the blood. The cure is simple. The application of fat or oil as soon as the tail becomes red and cold, twice per day, and continued for two or three days will almost always result in a cure. For some years we used boro-glyceride, a compound, we believe, of boracic acid and glycerine, but we are not certain that it is now procurable.

The third of the common troubles of the young pig is fits of an apoplectic and epileptic character. As a rule the shortest, thickest, and fattest pigs of the litter are those which are affected. This points to the chief cause, too much food in the form of mother's milk. A reduction in the quantity of the food fed to the sow will generally be effectual, except when the pigs are old enough to eat. Then both the quantity and the quality of the food given to them should be reduced. Prompt removal of the cause is usually sufficient, but it may be necessary to mix a little medicine in the food in persistent cases, or when remedial measures are not promptly taken. The ordinary symptoms are unmistakable, the pig falls on its side, struggles and gasps for breath, then in a minute or two it rises and appears to be little the worse. Unless continued over a period, fits are not usually fatal.

Very occasionally young pigs suffer from the protrusion of the rectum, or as it is commonly termed "shooting of the gut." This is due to various causes which result in straining. Of these constipation and diarrh[oe]a are the two most common. As soon as the protrusion of the gut is noticeable, the enlargement should be carefully washed, then oiled and gently pressed back into its natural position. Some pigmen advise the dusting of flour on to the protruded portion before it is returned, but there is a risk of increasing the amount of inflammation which is generally present. If known the original cause of the trouble should be removed, but in any case it is advisable not to give any solid food to the pig for two or three days after the operation.

Still another of the troubles to which pig flesh is heir is hernia, or rupture. This is of two kinds, umbilical and scrotum. The former is the escape of a portion of the bowels through an imperfectly closed navel opening, whilst the latter shows itself in an enlargement of the scrotum or purse due to an escape into it of a portion of the abdominal contents. Both of these ailments are considered to be hereditary, but the most common and the most troublesome is the latter, since there is always a chance of strangulation of the escaped portions, which nearly always results in death.

At one time it was considered to be inadvisable to castrate the boar pigs affected, but of late years the plan has been adopted of making only one incision in the scrotum in place of two, and making that one as high as possible. Then after the operation is performed, the aperture is sown up. The pig should be fed lightly for a day or two in order to give time for the healing of the wound.

Umbilical hernia is not generally of much importance, the navel opening gradually closes as the pig grows stronger and the enlargement disappears. It is advisable not to breed from a sow pig which has been affected, nor to continue to use a boar which has begotten ruptured pigs, as both failings are hereditary.

For a time at least, there is certain to be a difficulty in obtaining a full supply of sharps, even of the greatly reduced feeding value of the present quantity available. It may, therefore, be advisable to refer to another system of feeding the suckling sow and the young pigs. It is now perforce being generally adopted, but the result is not generally considered to be equal to the old system recommended. It is that of feeding pigs of the kind mentioned on vegetable food, and a mixture of palm nut, cocoa nut, ground nut, or linseed cake. The proportions fed at the Cambridge University Farm are mangolds 20 lbs.; a mixture of two parts palm nut cake, and one part cocoa nut cake, 2 lbs.; linseed cake, 2 lbs.; and ground nut cake, 1 lb. The two former were fed in the morning and evening, and the other two at midday. The various cakes seem to have been fed in a dry condition, but other pig feeders have found it beneficial to soak the cake in water for some twelve hours. This view seems to have received support from the practice at Cambridge, which was to mix the cake with the cut mangolds twenty-four hours before being fed to the pigs so that at least a portion of the cake would become softened by the mangold juice. Almost any kind of vegetable matter containing a fair amount of nutrition would be equally as suitable as mangold, indeed more so during the period from October to April. In the winter months cooked potatoes; kohl rabi, swedes, parsnips, cabbages, artichokes, etc., fed raw; and in the summer grass, lucerne, clover, vetches, rape, or almost any kind of vegetable food will be readily eaten by the pigs. Even where the wasteful practice of peeling the potatoes before being cooked for the household is still followed (and just how wasteful this old-fashioned plan is has been lately proved to be a loss of nearly one quarter of the nutriment)-- it is advisable to boil the parings and then mix the whole with the pig's food.

It cannot be too strongly impressed on pig keepers that a certain proportion of vegetable food is most beneficial for pigs of all ages, as not only is a saving in cost effected, but the pigs will continue in a more healthy condition than when fed solely on meal or other concentrated food.

* * * * *

[Illustration: LARGE WHITE BOAR. The property of the Author. The Winner of many Prizes.

To face page 80.]

* * * * *

[Illustration: Photo, Sport and General.

TAMWORTH SOW, "QUEEN OF THE FAIRIES."]

CHAPTER X

WEANING PIGS

There are few points in connection with the breeding and feeding of pigs on which there is a greater diversity of opinion and practice than on the question of the weaning of the young pigs.

For instance, take the age at which it is most satisfactory to take the pigs off the sow. This practice varies greatly in different districts, and even in the same district where one would naturally suppose that the determining influences would be similar weaning at five or six weeks old.

One pig breeder will declare that a little pig of five or six weeks old should be and is able to support itself alone, and will act accordingly. Should perchance a litter weaned thus early cease to grow the excuses made will be various.

The weather is at fault, it is either too hot or too cold, or the sharps, etc., on which they have been fed were not good or sweet, that the sow's milk was not sufficiently plentiful, or it was wanting in nutriment. In fact, any excuse will be made rather than the actual cause admitted.

In far too many instances the real reason for the want of thrift on the part of the young pigs taken from their mother when they are not more than five or six weeks old is that their digestive organs are not sufficiently developed as to enable them to digest enough food to nourish them properly.

Another excuse often made for what we consider to be undue haste in weaning young pigs is the alleged desire of the owner not to waste the time of the sow. He is anxious to have her served again and hasten the arrival of the next litter.

Occasionally it is found to be unnecessary to wean the pigs for this purpose as the sow will come in heat and can be served by the boar, but if she should become in pig the result will be much the same so far as the pigs are concerned, since as soon as the sow has conceived the milk will promptly cease or become very reduced in quantity and quality.

On the other hand, if the sow does not stand to the boar time may be wasted. It is most unlikely that the sow will again become in heat for some three weeks, whereas this almost always occurs within a few days of the weaning of the litter of pigs.

Then another extreme, and one which is practised by some pig breeders, is to allow the young pigs to remain on the sow until the former are from ten to twelve weeks old. It is claimed for this practice that the young pigs grow much faster when left on the sow than when weaned, and that less food is consumed for a live weight increase from a given quantity of food. Also, it is said that food of more inferior kind can be fed to the sow than could be fed to the pigs if they were weaned, and thus the sow and litter are kept at less expense, and that if the pigs are not weaned until nearly three months, the milk of the sow will have gradually ceased to flow, and the pigs will not miss the help from their dam. Their digestive organs will then have become sufficiently developed to enable them to make the best use of the food given to them, and they will sustain no check in thrift or growth when they are weaned.

In this question of weaning pigs the good old fashioned plan of following the middle course will probably be found to be the best. Anyway, it was the one which we followed for a great number of years and found the results generally satisfactory for the following among other reasons.

As a breeder of pure bred pigs for sale as boars or yelts for breeding purposes, we were naturally anxious to give the pigs a good start in life so that we should be able to sell them as quickly as possible, and that they should thrive when they came into the possession of their new owners, and thus prove the best possible advertisement of our herds. As a rule we found that if the pigs were allowed to remain on the sows until they were some eight weeks old they were quite strong enough to fend for themselves, that by gradually increasing the length of time which the sow was allowed to remain from the pigs, the latter became accustomed to exist without the mother's milk, and as the milk of the sow naturally dried up when the pigs partially ceased to withdraw it, no trouble was experienced with inflamed udders as is usually or commonly the case when the pigs are suddenly weaned from a sow which is in full milk.

There is also another advantage apart from that to the sow and pigs, it is that the sow will almost invariably come in heat within three or four days of the weaning, and with the best possible chance of becoming in pig.

Some pig keepers are more inclined to wean their litter of pigs at an early age, and then if the sow be low in condition to baulk her at the first time of [oe]strum. There are objections to this--one of them is that there is frequently a difficulty in getting the sow to conceive after she has been baulked. Why this should be so we have not been able to ascertain. We only record what we know to be a fact.

In our opinion this difficulty is one of the strongest points in favour of the practice of allowing the young pigs of a sow with her first litter, or of an old sow which has become low in condition (either from having had too many pigs left on her, or from other natural cause), to remain on the sow for a longer period than about eight weeks. Some persons will keep the pigs on the sow until they are nearly three months old in the belief that both sow and pigs are benefited, and that the pigs can be kept quite as cheaply if not more so when unweaned than weaned. They also claim that the sow is so much stronger and better fitted to prepare for another litter. Experiments have been carried out in the United States which go far to prove that the first of these two claims is founded on fact; and it has further been demonstrated that certain foods can be fed to the sow without affecting the thrift and health of the pigs which could not with safety be fed to the latter direct, yet when fed through the sow the pigs will thrive on the milk produced therefrom. It is entirely a question of the cost of a rest for the sow during the extra two or three weeks, and the benefit to the sow and her pigs.

One occasionally sees in the press a claim for what is considered to be a great achievement in that some one has bred three litters of pigs from one sow within the year. There really is something wonderful in this since of the fifty-two weeks constituting a year, the sow would be carrying her pigs some forty-eight weeks. This would allow only four weeks for the two litters of pigs to be suckled, and this would also include the few days between the pigs being weaned and the sow coming in heat. Apart from the natural difficulty of successfully breeding three litters of pigs from one sow within twelve months, there exists a far greater possibility of loss rather than of gain from unduly hurrying on the arrival of each litter of pigs from a sow, especially of the profitable kind of sow.

Some forty years since when Small Whites, Small Blacks, and short thick Berkshires were fashionable, the number of pigs in each litter was few, and the number reared still fewer, owing to the limited quantity of milk furnished by the sow. Now, the Large Black, the Large White, the Middle White, the Lincolnshire Curly Coat, the old Gloucester Spots, the Tamworth, the Cumberland, and even the sows of most of the local breeds of pigs are expected to rear nine or ten pigs each litter. Even if it were possible for a sow to bring forth three litters within the year, she could not possibly do justice to them either before or after the piglings arrived in this world; and further, the life of such a sow would of necessity be a short one. It must not be forgotten that in the production of each litter of pigs the sow is compelled to manufacture from 20 to 30 lbs. of flesh, skin, hair, etc., which together constitute the newly farrowed pig, and very frequently this has to be accomplished on a far too limited supply of suitable food.

* * * * *

[Illustration: From a Painting by Wippell.

MIDDLE WHITE SOW.

To face page 96.]

* * * * *

[Illustration: Block kindly supplied by Cumberland Pig Breeders' Association, Carlisle.

CUMBERLAND SOW.

Owned by Mr. Carr, Kirkbride, Carlisle.]

CHAPTER XI

THE REARING OF YOUNG PIGS

One of the most important points in the profitable raising of stock is to give the animals a good start in their earliest days. There is an old and true saying amongst shepherds that the best and most profitable sheep are those which have never lost their lamb fat. It may with equal truth be declared that the most profitable pig is the one which has a good start when on its mother, and never afterwards lacks suitable food, judiciously given.

At frequent intervals, the question as to the number of pigs which a sow should have left on her to rear is the subject of discussion in the press. At each of these periods very similar arguments for and against large litters are used with much the same inconclusive results. This probably arises to a great extent from the varying conditions under which the particular litter of pigs is to be reared. The time of year has a vast influence, a sow farrowing in May will more successfully rear a dozen pigs than she would bring up ten if they were farrowed in the month of October.

The age and condition of the sow should also be considered. A young sow of about twelve months should not have more than seven or eight pigs left on her to rear, whereas with her succeeding litters until she is at least four years old she would rear at least ten pigs each litter. After the sow has reached the age of about four years, if that time has been fully employed in her maternal duties, she becomes gradually less able to rear so large a number of pigs with an equal amount of success. It is then advisable to vary the number according to the season, and to the physical condition of the sow; generally speaking an aged sow will bring up more pigs in the summer months than in the colder months.

Reference is made elsewhere to the manner in which the young pigs should be cared for until they are weaned from their dam, but no harm can arise from a repetition of the advice that the young pigs should be so managed and fed that only the very slightest difference will be noticed by the youngsters when parted from their mother. Many pigs are permanently checked in growth by being suddenly deprived of a full supply of mother's milk if weaned when their digestive organs are insufficiently developed as to treat a sufficiency of food to make growth and progress without the assistance received from their mother's milk.

Opinions differ as to the age at which little pigs are sufficiently developed as to exist and thrive without their dam's help. Here again the time of the year, not only as far as the weather is concerned, but the desirability of prompt or deferred remating of the sow in order that her succeeding litters should arrive during the most favourable portions of the year, must be considered. The thrift and growth of each litter of pigs varies greatly. The health of the sow, her condition before farrowing, and other causes, some of which are not always on the surface, have their influence, but it may be taken as a rule that young pigs are fully able to fend for themselves by the time they are eight weeks old. Should it be possible to allow the pigs to remain on the sow for so long a time as twelve weeks without prejudicing the next litters as to the time of year of their arrival, the pigs may benefit, and no loss of food will be sustained, since it has been clearly proved that pigs beyond the age of eight weeks can be fed quite as economically, if not more so, on the sow than if weaned. It may also be possible to feed the sow on somewhat coarser and less expensive food than could be satisfactorily fed to the young pigs, as her digestive organs would be better able to treat the coarser food. Another advantage generally follows keeping the pigs for a longer time on the sow if the latter be well fed is that she will be in a stronger and better condition to start the building up of her next litter.

As a rule young pigs will commence to eat when they are from three to four weeks old. If the sow is fed in the sty in which the little pigs are, these will endeavour to share in the food; at first they may content themselves with licking any food which may be outside the trough, but they quickly show a desire for more, and attempt to get into the trough. When this is evident, it is advisable to feed the little pigs apart from the sow; a low flat trough is best, as one with high sides is said to cause "high backed" pigs, or pigs suffering from a curvature of the spine. If a little milk can be obtained, the pigs will promptly drink it, if the milk be whole they will thrive best, but even if only skim or separated milk be obtainable, or butter milk, providing that it be drawn off ere the salt is put into the churn, a small quantity will be beneficial, but the pigs will not be able to digest so large a quantity of the separated as of the whole milk. The former is apt to have a constipating effect on the bowels of the youngsters. Should an ample supply of separated milk be available it can be fed through the sow, who will be better able to digest it, and whose yield of milk will be increased, provided that sufficient separated milk to affect her bowels be not given to her. A few kernels of wheat or white peas will be readily eaten by the little pigs, which will benefit therefrom.

If no other food is available, sharps, or whatever the local term for the finer miller's offals may be, mixed with a little warm water and fed to the piglings, will prove beneficial, care being taken to give only so much as the pigs will eat up readily, or that any surplus is taken away, so that it does not become sour, as in this last condition it will cause diarrh[oe]a in the young pigs.

When the pigs are about six weeks old the sow can be allowed to remain from them for a longer time, and the youngsters fed two or three times each day. The sow's milk will then gradually dry up, and the pigs will become accustomed to the food, so that when the latter are about eight weeks old they will have become weaned naturally, and receive no check from the loss of the sow's milk. This system, will also prevent any trouble arising from the collection of milk in the sow's udder, and the occasional attacks of inflammation or garget which follow a chill to the sow when her udder is in an inflamed condition from being closely impacted with milk.

Assuming that the economical and beneficial practice of supplying the suckling sow with vegetable food of some kind after the pigs are some three or four weeks old has been adopted, the pigs will have become accustomed to its consumption. It will be found to be advisable to continue this whether it has taken the form of cooked potatoes, of mangolds, swedes, kohl rabi, cabbages, artichokes, etc., as not only will the food bill be reduced, as the pigs will make equal growth and thrift on food containing say ten per cent of vegetable matter as they will if fed wholly on sharps, but the vegetable food will have a beneficial effect on the health of the pigs, and tend to prevent those attacks of constipation and diarrh[oe]a which are so frequently the result of food of too rich a character.

Of the vegetable foods, cooked potatoes and raw artichokes are the most nourishing and the most readily eaten, lucerne and clover in a green state come next in food value and favouritism with the pigs; cabbages are credited with causing constipation when fed to young pigs, whilst mangolds are said to have the opposite effect, and in addition when grown on light land by the aid of artificial manure mangolds are apt to affect the kidneys and cause excessive urination. Kohl rabi are not so much used in the feeding of pigs as would be advisable. They are easily grown and will take the place of swedes on land on which swedes are subject to mildew; they are very nutritious, and are readily eaten by both old and young pigs.

Coleseed is not used in the feeding of pigs in this country to anything approaching the same extent as in Canada and the United States; its value and results are of a very similar character to those of cabbages. Tares or vetches contain too large a proportion of water for young pigs, and they also have a tendency to cause looseness of the bowels. The growth of maize for feeding to pigs in a green state has been recommended by some writers, but in practice we found it most unsuitable for young pigs, and of little value for aged pigs, owing to the small proportion of nourishment contained in it in comparison with its bulk. Further, pigs both old and young will refuse to eat it unless driven by hunger. It is needless to remark that no pigs, especially young ones, will thrive under such conditions.

One of the most common mistakes made by pig feeders is allowing too long a time to pass between feeding times. Twice or three times per day is considered to be quite frequent enough, whereas prior to their being weaned the pigs would have had a meal wellnigh each two hours both day and night. Infrequent meals result in the pigs becoming so hungry that they bolt their food, and a greater quantity than is desirable, and then suffer from indigestion.

It must also be remembered that the pig's capacity for storing food is very small, especially as compared with some others of our domesticated animals. Four or five meals per day at least should be given to newly weaned pigs. That most troublesome of ailments commonly termed cramp more generally results from injudicious feeding than from all other causes combined. Even when the young pigs are properly fed on suitable food there is a tendency in some little pigs to attacks of cramp. One of the best preventatives and even remedies is to compel the pigs to leave their nests late in the evening or prior to the pigman retiring for the night, as they will then relieve the bowels and bladder. Otherwise, particularly in cold weather, the pigs would remain quiescent in their nests from feeding time in the afternoon until they were fed the following morning, or in winter a period of some fifteen or sixteen hours--far too long a time for the good health of the young pigs.

Another point which requires attention is the provision of a dry bed. Pigs are naturally clean animals, and will not as a rule foul their bed when they are in a healthy state. Still the straw will in winter time become damp solely from the moisture thrown off by the pigs when huddled together in their nest. All damp litter should be carefully removed at least once each day.

The best of all materials for the bedding of pigs is wheat straw. This will absorb a larger amount of moisture than any other kind of straw, whilst the skin and hair of the pigs will remain of a brighter colour than if bedded on oat or barley straw. Of these two, the former is more suitable than the latter, which so readily becomes damp and foul. In those parts of the country where comparatively little corn is grown, sawdust and wood shavings are commonly used for litter for pigs. So far as the comfort of the pigs is concerned there is little difference as compared with straw with regard to pigs of all ages in the warmer weather, but in the winter little pigs suffer, as they are unable to make the warm nest which straw enables them to make and enjoy.

When peat moss was first introduced it was strongly recommended for the bedding of pigs. It was claimed for it that it was a far better absorbent of moisture than sawdust, and that its manurial value was much greater. It is probable that both claims are founded on fact, as sawdust is of comparatively no value as a manure. But there exists one serious objection to the use of peat moss as litter for young pigs. It is that the pigs are given to eat it, that it causes severe attacks of indigestion, and often the death of the pig eating it.

Of late years the spaying of the sow pigs has ceased to be general. The causes of this neglect may be several, amongst them the dislike of trouble, but perhaps the main reason is that the so-called store period of the pig's life is now so much shorter than in the olden days, and consequently the loss of food, and the risk of the arrival of unexpected litters of pigs are less, from the repeated periods of heat, indeed under the present or recent conditions of pig keeping a large proportion of the pigs are killed ere they have become sufficiently developed to be troublesome in this respect.

Still, there is little doubt that the castrating and spaying of young pigs at about the age of six weeks, or before they have been weaned from the sow is advisable and the cost of the operation is well repaid. An unspayed sow pig becomes a nuisance in company with other pigs, and when it is put up to fatten will make no progress on some three or four days during each three weeks when she ordinarily becomes in heat.

In addition to her own waste of time she will, if penned with others, be continually worrying her mates and preventing them from resting and thriving.

Until recently another objection was taken to the unspayed sow pig, it was that if she were killed during the period of [oe]strum that great difficulty would be experienced in curing the meat properly, and that signs of her heated condition would be noticed in the mammary glands in the form of dark globules of what was considered to be blood, but investigation carried out at the University Farm at Cambridge by Messrs. Russell and Kenneth Mackensie have proved that the discoloration and the consequent loss in value of a certain portion of the belly of a side of bacon is not due to the pig having been in a state of heat at the time of its slaughter, but to an excess of pigment, noticeable only amongst coloured pigs. Thus, the globules would be of a dark colour when the bacon was from a pig of a black colour, and red from the pigs of the Tamworth breed. This shows another cause of the marked preference of the bacon curers for pigs of a white colour in the manufacture of the highest priced bacon.

CHAPTER XII

HOUSING OF PIGS

In the general management of pigs there are many points on which improvements might be effected without any very considerable amount of trouble or expense. Far too frequently this neglect or want of care and thought is observable in the housing of pigs. Many of the sties in the country districts are neither wind nor water tight, and they are far too often in a most unsanitary condition, indeed in such a disgraceful state that some excuse was afforded for the drastic, if injudicious order of the sanitary authorities which prohibited the erection of a pigsty within from sixty to one hundred feet of a dwelling house. Undoubtedly it would have been wiser to have permitted the keeping of pigs within a much shorter distance of the house only so long as the necessary steps were taken to prevent a nuisance or a risk of the residents in the house suffering in health. The proximity of a pigsty to a house can be rendered perfectly innocuous with ordinary care, and the cottager not be deprived of very considerable advantages not only in making a profit, but in the provision of manure for his allotment or garden which will benefit greatly from its application.

The mistakes or want of care in the erection of pigsties is by no means confined to the owners of cottages or small holdings, as a considerable proportion of the piggeries on which great outlay is expended are equally as unsuitable if not so insanitary. Even in so-called model buildings the piggery has often been the last thing thought of; the stables, the cow house, etc., have been conveniently placed for feeding the occupants, for air, light, and sun, and then the piggery has been placed in whatever spot may have been left unoccupied, and as this generally happens to be on the northern side of the buildings, the unhappy pigs are deprived of the rays of the sun, which are to them quite as necessary, if not more so, than to any others of our domesticated animals.

This same want of sun, and the exposure to cold is noticeable in only a lesser degree in those buildings which comprise a double row of sties with a passage down the centre, a store and a cooking and mixing house at one end, and an exercise or feeding yard adjoining. It matters not whether the building be placed north or south, or east or west, one half of the sties have a wrong aspect; even if the sties facing the west can be said to possess one. The trouble is still greater with the system of having a yard attached to each sty. The north or east wind renders the sties with such an aspect a most uncomfortable and unhealthy place for young pigs during more than half the year, whilst older pigs cannot thrive on the same amount of food as they would if their quarters were comfortable. Apart from the waste of food which results from these draughty and cold sties, the latter are the chief cause, with injudicious feeding, of that most troublesome ailment amongst pigs, rheumatic gout, or, as it is commonly termed, cramp. How very draughty and uncomfortable these sties are which have an open yard attached, and an inlet at all times usable, can be readily discovered in cold and windy weather by noticing the position in which the occupant has made its bed. This will be found not on the highest part of the sty, which will be opposite to the opening into the yard, but in the corner next to the opening, since in this position it is less exposed to the cold wind which rushes into the sty through the opening.

Apart from the unhealthiness to the pigs resulting from the exposure to draughts it is not apparent to the writer that any advantage is gained from the provision of these yards. In many instances they serve only for an excuse to limit the height of the sties, as unless these are of a fair height there is a considerable difficulty in cleaning them out. The money expended on building the yard would easily cover the extra cost of raising the side walls of the pigsty by two feet, and thus not only render it free from draughts, but also make it far more healthy and less subject to the extremes of heat and cold.

The ordinary sty with a yard attached is unhealthy for a growing or matured pig, but in the colder weather it is simply cruel for newly born pigs, of which numbers are annually lost from exposure or are greatly checked in their growth.

One of the very best places in which to house pigs in the experience of the writer was a large barn with a thatched roof. This was divided off into sties by partitions some 4 ft. 6 in. high; owing to the height of the building the temperature was not unduly high in the hottest weather nor did the pigs suffer to any extent during severe weather. These advantages arose mainly from the slight changes in temperature, and an abundant supply of uncontaminated air.

One of the greatest drawbacks to the majority of the pigsties is the absence of ventilation without draught. This trouble is especially noticeable where the side walls are not more than about 4 ft. high, whilst the proximity of the roof to the pigs increases the sufferings of the pigs from the heat when the weather is excessively hot.

Some of our most successful pig feeders on a large scale have found it profitable to erect cheap buildings very similar to small barns, the side walls being at least 10 feet high. This will permit of thorough ventilation, quite free from draughts, whilst the variations in the temperature will be comparatively slight. The building being complete within itself, and entirely used for the pigs, there is no disturbance of the pigs between the feeding times, so that the pigs will rest and grow fat. These houses are most suitable for a number of fattening pigs, whereas for sows and for young sows smaller sties or houses are more convenient. These should be at least 10 ft. square, the front 6 ft. 6 in. high, the doors divided so that the upper half can be opened when the weather is favourable; ventilation can be obtained by hanging or sliding doors just under the eaves so that the pigs are not affected by the draught; the floor should be laid with brick and gradually incline to the front of the building so that the liquid can run through an aperture in the lower part of the front wall into a cesspool placed close to the building. A row of these houses, which should face to the south, can be more cheaply erected than a single house, as the wooden partitions between the houses need not be more than 4 ft. high, and one of these would take the place of two gables or ends. Several of the houses which the writer erected had brick foundations and feather-edged boarded sides and ends; the roofs were of tiles unpointed, as in this way the houses were much cooler in the summer, whilst in the winter the upper portions of the houses were packed with straw which still permitted of the escape of the foul air, yet greatly added to the warmth and comfort of the building.

The one thing of all others most needful in the sty or house for the well doing of pigs is a sufficiency of pure air without draughts; pigs of even a few days old will suffer less from cold than from moist and foul hot air. It is not the most costly building in which pigs will thrive best, but the one in which they are the most comfortable and free from the extremes of heat and cold with a dry bed on which to rest and be thankful.

When making a tour of the Agricultural Experiment Stations and Agricultural Schools in Denmark some few years since, the writer saw near Aarhuss what was then a novelty in the form of a two decker pigsty, i.e. a sty with a sleeping place above--one could scarcely term it an upstairs room as access was gained not by stairs but by an inclined board with struts of wood fastened across it to give a firm holding to the pigs as they ascended to the upper story. The incline was very steep, but the pigs seemed to have no difficulty in getting up and down. The advantages claimed for it by the principal were that the sleeping compartment was so much cleaner and sweeter; that less straw was required for bedding, and that the pigs were far more comfortable and rested better than when boxed up, especially in the summer season when the heat in the lower portion was very oppressive. The feeding took place in the lower portion. It was stated that nearly the whole of the urine and dung was deposited below. This was a great advantage as the moisture ran off at once into the drains, and the solids were easily cleared out as there was no litter mixed with them, or the dung could be readily washed into the drains by water from a hose, which was used in the summer for the purpose of bathing or of washing the pigs.

The chief objection to the plan would be its expense, as unless the pigsties were in a barn or a shed already erected for some other purpose the pigsty would have to be so much higher on the side walls and consequently more strongly built.

* * * * *

[Illustration: Photo, Sport and General.

LARGE WHITE SOW, "WORSLEY SUNBEAM."

To face page 112.]

* * * * *

[Illustration: Photo kindly lent by Kenneth MacRae, R.U.A.S., Balmoral, Belfast.

LARGE WHITE ULSTER BOAR.]

CHAPTER XIII

THE EXHIBITION OF PIGS

When the exhibition of live stock at our numerous shows became common, a belief sprang up amongst non-exhibitors that the preparation for show was most deleterious to the animals shown. It was also contended that exhibitors were prone to pay attention, to a far greater extent, to the fancy or show points of the animals which they bred than to those utility points which are of infinitely more importance to the ordinary stock breeder and the consumer. It was also believed that the feeding or training which the show stock underwent seriously affected their procreative powers, and especially so with the animals of the feminine gender.

It may at once be frankly admitted that there existed some ground for the belief that a majority of the exhibitors did appear to give too great attention to the claims of the judges who were, in too many cases, chosen for reasons other than their knowledge of practical agriculture or the requirements of the consumers of meat. For so acting, the exhibitors were not beyond blame, as in the earlier days of showing, their main object was to win prizes in order to advertise their stock and so secure customers for their spare breeding animals. The actual improvement of the various breeds of stock did not in those far-off days appear to be of such vital importance as the world upheaval, of which the present generation has been the witness, has proved it to be.

It may also be fairly claimed that there has been some slight improvement in the system of feeding and training followed by the pig exhibitors of to-day. This is in part due to the fact that the cramming on rich food and giving little exercise may result in rendering the show pig in such a state of obesity as to secure the approval of the non-practical judge, who is unable to appraise the points of a pig when in its natural breeding condition, but that to be able to follow the present system of exhibiting at several successive shows and even when the bloated pig is intended to be returned to the breeding pen, this excessive feeding proved to be a grievous mistake. It may not be possible to claim that the over feeding of show animals is a thing of the past, but there is little doubt that exhibitors of pigs have become alive to the fact that it is not profitable. Not only is the expense excessive, but the damage done to the breeding animals is so great as to render it inadvisable for any ordinary farmer to follow. Again, there has of late years been a very considerable improvement in the pig classification at both the breeding and the fat stock shows. When the writer began pig showing, on his own account, fifty years since, the common classification at most of the shows was, boar any age, sows any age, and pens of three breeding pigs, not exceeding nine or in some cases even twelve months. There were no restrictions as to the age of the boar or of the sow, no condition as to utility, of the sow having at any time reared a litter of pigs or of being in pig, so that it was by no means uncommon at even some of the chief shows to find both boars and sows appear year after year, having been guiltless of any attempt to procreate their species, but having been kept solely for the purpose of winning prizes and adding to the renown of their owners, if not directly adding much to their balances at the bank. The only way in which the continued exhibition of these old stagers was made profitable was the securing of customers for breeding stock from the exhibitors, who in far too many cases were not the breeders of the winning animals. To so great an extent had this purchase, frequently from middlemen or dealers of exhibition pigs, become in the seventies of the last century, that some of the live stock papers in the United States took up the cudgels on behalf of the American breeders of pigs, who had been in the habit of importing show winners from this country and plainly asked for the English definition of a pure bred pig. It was pointed out at a recent show of the Royal Agricultural Society several winners shown by one exhibitor were entered as of certain defined breeds, yet neither age, pedigree, nor name of breeder was given, the only particulars given in the show catalogue being the name and address of the exhibitor, the name of the pig, and the further statement age and breeder unknown. As our American cousins asked, how could it be possible to ensure that a pig was of a certain pure breed when it was admitted that no knowledge existed of the breeding of the animal nor actually of the person who bred it. This scandal, as it was termed, was one of the contributing causes of the establishment of societies for the registration of the pedigrees of the various types or breeds of pigs.

Other changes which have been great improvements have been the limitation of the ages of boars and sows shown, the requirement that the sow has within a certain fixed time farrowed a litter of pigs and that when entered as being in farrow a certificate of subsequent farrowing shall be furnished ere the prize money is paid over. The age of the young boars and sows has also been reduced at most shows to six months, or the pigs must have been farrowed in the year of the show. In the good old times the age of the pigs shown in the classes for pens of two or three or five, varied from six to twelve months, and the asserted age given by the exhibitor was accepted as correct. At many of the important shows not only are some means of identification asked for, but the state of the dentition are variously dealt with; at some shows they are disqualified at once by the stewards on the certificate of the veterinary surgeon. It may at once be admitted that this mode of procedure is very hard on an honest exhibitor whose pig has for some reason developed its temporary or permanent teeth abnormally--and such cases are not unknown--- although as a rule the various stages in the cutting of the permanent teeth are very regular, the majority of the irregularities are also in favour of the exhibitor, since delayed rather than precocious development of dentition is the most common. Just how imperative it was that some steps should be taken to prevent mistakes being made in the ages of young pigs exhibited, many cases could be cited, but one may suffice where one of the sow pigs in a pen of five entered in a class for pigs not exceeding six months actually farrowed a litter of fully developed pigs in the show yard.

During the last forty years, great improvements have been made in the classification for pigs at our principal Fat Stock shows. The division of breeds or types has been attended to and the ages of the pigs in the various classes have been greatly reduced. For instance, when the writer was judging pigs with two colleagues at the 1880 show of the Smithfield Club, there were classes for Small White pigs, not exceeding nine months; above nine months, and not exceeding twelve months, and above twelve months and not exceeding eighteen months. A more ridiculous classification could not possibly have been devised since no small white pig would have paid for fattening after it had become nine months old. A similar classification existed for pigs of the Large White breed, for Black breeds, and for Berkshires. In addition there was a class for a single pig of any age or breed. The condition of some of the exhibits in the oldest classes was most pitiable, they had been stuffed to such an extent that their life must have been a misery to them, they were unable to walk any distance, and to prevent suffocation rollers were used on which to raise their heads. The only way in which to describe these unfortunate subjects of man's inhumanity was as animated bladders of lard.

At the recent shows of the Smithfield Club, not only has the age limit been greatly reduced but classes for pigs not exceeding 100 lbs. live weight have been instituted, in addition to classes for all the recognised pure breeds of pigs and those of any cross. Even this great reduction in age has not been enough to satisfy some of our reformers, as an endeavour is being made to reduce the limit of twelve months to nine months, so that in future the classes will be for pens of two pigs not exceeding 100 lbs. live weight, for pigs not exceeding six months old and for pigs between six and nine months old, with certain classes for single pigs under nine months. It is contended that fat pigs cannot be profitably kept after they reach the age of nine months. Another innovation of recent years at the Smithfield Show has been the establishment of the so-called slaughter classes. This is probably by far the greatest improvement of recent years in the pig section. Classes are provided for pigs not exceeding 100 lbs. live weight, pigs weighing over 100 lbs. and not exceeding 220 lbs., and for pigs above 220 lbs. and not exceeding 300 lbs.

live weight. The pigs are first exhibited and judged alive, then slaughtered and the carcases judged on their pork merits. There is also one class for pigs above 160 lbs. and not exceeding 240 lbs. live weight best suited for the manufacture of bacon. These various classes have created great interest and have proved of the greatest educational value.

Another beneficial effect of the changed conditions is the elimination from the summer show-yards of fat sows guiltless of milk and accompanied in the pen by half a score of young boars and yelts of an age varying from three months upwards, and which together were exhibited in the class for breeding sows, or breeding sows and pigs. A fine fat sow which would take kindly to an unlimited number of adopted youngsters was in those days almost as valuable as a small gold mine. An old and well-known pigman, Dick by name, assured the writer that no fewer than sixty-three young boars and yelts were sold in one year off or when in company of one well-known sow. At the present time the pigs shown with a sow must be certified to be her produce and not to exceed the age of eight weeks.

It is at all times difficult to discover the motive power for certain actions on the part of a human being. It has been declared that there is an equal amount of doubt as to the cause of a breeder of stock wishing to exhibit his animals. Surely this last assertion is at least of a doubtful character. What greater proof could a stock breeder give of his pride in his animals than a burning desire to expose their good qualities to the public gaze. In addition to this, few men are entirely free from the spirit of gambling and this enters into all competitions, particularly in the show yards. The winning of prizes with stock may not be quite so uncertain as the winning of horse races, still, there is enough of uncertainty to render the judging ring a centre of great excitement. Some persons will even contend that the showing of farm stock is not desirable on the part of young farmers as it is likely to assume so great a similarity with gambling, that attending the shows means a neglect of business and leads to expensive habits. On the other hand, it cannot be denied that the exhibition of our improved specimens of stock has been of untold benefit to both home and foreign stock breeders. Further, the exportation of our pedigree stock has actually saved us from semi-starvation during this most fearful of all wars, as without our improved stock the native stock of foreign countries could not possibly have furnished the enormous quantities of meat which we have had to import.

It may be that a great many exhibitors of stock had little or no intention of becoming one when they first purchased their stock, but on these proving quite the equal of that possessed by their neighbour, the desire grew to suggest how good they were, or in many instances the original entries have been made in response to a request to support the local show.

This may be still another cause for a beginner in stock breeding exercising extreme care in the selection of his original stock. Even if the prime cost be higher than that of ordinary market stock the extra outlay expended on animals from well-known breeders, and out of old established herds, is certain to prove a good investment. There is just as great difference in the different families or strains of our domestic stock, as there is in the various human families and of animals, and it may be probably more true that the vast majority of the best of them are the descendants of a comparatively few ancestors. This is evident in almost everyone of the breeds of our improved stock, it is so in thoroughbred and shire horses, and so one might go through the whole list of domesticated or farm animals.

It is therefore desirable that anyone who thinks of exhibiting his pigs should endeavour first to discover the particular tribes or families which, in the past, have furnished a large proportion of the winners, and then to obtain some of the specimens of those families which have been successful in the show yards and in the breeding pen. This combination is most important, as it does not necessarily follow that a line of blood which produces prize winners shall also produce animals which are not only good in type, character, and form, but possessed of prolificacy, free milking properties, and ability to raise large litters. The difficulty of finding in some of the mere exhibition herds this most desirable combination is due, in the main, to the far too frequent neglect of the utility points, the two aims of the herdsman are in too many instances the winning of prizes for their employers and the securing of a percentage of the prize money for themselves.

Although there have been attempts made to impress on outsiders the claim that there exists in the training of pigs for successful exhibition in our show yards a large amount of mystery, yet, the practice is most simple, it consists in the employment of the greatest possible observation, care, and attention; without the continual use of these qualities it is not possible to become a really successful pigman. In very many instances just that little extra attention has turned the scales. The one chief qualification on the part of a successful stock man is the art of taking pains. Unlike most of the other exhibitors of pigs who exhibited largely over many years the writer never employed a professional pigman. The comparatively small number of pigmen who assisted him to win thousands of prizes were merely ordinary farm labourers, save in one case, and he was an old sailor, yet one of the best feeders and trainers we ever employed. He was naturally fond of animals and was never tired of waiting on them and of supplying their needs. It was once jokingly said of him that, having no children, he bestowed on the pigs in his care the love which some other people bestowed on their children. There is much of truth in the assertion made by a coloured preacher in the United States when discussing the want of success of ordinary pig-keeping in the States, the chief cause he declared was the absence of love. We would call it want of natural fondness of animals and an insufficient determination to render the conditions of life of the animals in our charge as pleasant and satisfactory as circumstances will allow. With regard to the system of rearing and feeding animals intended for exhibition, nothing more is needed than the concentrated care and attention which is required in the successful rearing and feeding of all commercial animals. A liberal supply of suitable food, prepared in the most tempting form and judiciously fed to the pigs in just the quantity required, as frequently as the pig is able to thoroughly enjoy it. Little and often is a good motto for the pig feeder. The more closely we adhere to nature, the more successful shall we be. It is to this, perhaps, that exercise is so specially necessary for pigs which are being prepared for the show yard. It is impossible to render a pig perfectly fit for exhibition at a show, and more particularly at several successive shows, without plenty of exercise. Each morning and evening a walk of a distance varying with the ages, etc., of the pigs is desirable. Another point to which some professional pigmen give great prominence is the regular dosing of their charges with secret medicines. This is not only unnecessary, but may with breeding animals prove harmful. A sound healthy pig seldom requires medicine if it is properly fed and exercised. It is the over feeding or intense desire of the pigman which in the majority of cases renders medicine necessary.

A word of warning against this haste to get the pig into show condition. This last can only be a work of time, and the commencement of the process must be in the early stages of the life of the pig and be steadily continued until within a few days of the show. This slight reduction of the food may be necessary in the summer when the heat is great and the pigs become restless when travelling boxed up in a crate in an enclosed truck. Many of the pigs lost in travelling to or from the shows or soon after arriving at the shows, have been fed just prior to being loaded up, because of the difficulty in feeding them when on the journey. This is an entire mistake; not only should the pigs not be fed, but prior to being put into the crates they should be given just so much exercise as will cause them to evacuate the bowels, or the bladder. Care in this respect and non-exposure to the rays of the sun may not in every case prevent trouble, but it will most certainly reduce to a minimum the chance of it. Should a pig suffer from the heat, cold water should be applied to the head by means of a sponge or a cloth, and should some of the water percolate into the mouth of the pig so much the better.

CHAPTER XIV

PRESENT AND FUTURE PIG-KEEPING

As it is impossible to foretell the effect which the present disastrous war will have upon the pig-breeding industry, we have deemed it expedient to refer as briefly as possible to the present conditions of feeding, etc., which may or may not prove to be of a temporary character or which may become permanent in a more or less modified manner.

One of the results of the scarcity and high market value of the different articles which have been commonly used in the feeding of pigs is drawing greatly increased attention to the original conditions under which pigs were kept, i.e. when they were in a wild state or when they were allowed their partial freedom for the purpose of getting their own living to a greater or lesser extent.

We are aware that a claim has been made by an enthusiastic convert to pig-keeping that in allowing his pigs their liberty to roam over grass fields and in woods he is practising quite a novel course of procedure, but the old hands merely smile and admire the enthusiasm which is more nearly allied with youth than old age. The practice may not have been generally followed of late years, but in the middle of the last century it was to the writer's knowledge common in certain of the Eastern Counties, particularly in Suffolk and portions of Essex and Cambridgeshire, where a considerable acreage of grass and especially clovers was grazed by pigs, having a greater or lesser quantity of other food as the pigs were intended for breeding or fattening purposes.

Generally speaking, some shelter of a temporary character was provided failing that furnished by trees, and straw stacks, etc., but our American cousins have gone one better in that they have introduced small movable houses which can be transported on wheels and can be utilised for a sow and her pigs, or for a number of stores. In the former course, an enclosure sufficiently large for the sow to graze therein is fenced in so that each sow can be kept separate until the pigs are old enough to prevent others from robbing them of their birthright. The chief difficulty attending this system is not experienced in the United States to the extent it is in this country, since the general custom there is to allow each sow to farrow a litter of pigs in the spring and then to fatten off both sow and pigs, save those reserved for breeding purposes next year. This plan, which appears to be wasteful, also handicaps the owner who desires to improve his pig stock, since an opportunity is denied him of discovering the best of his sows and so reserving them and their produce to form the nucleus of a really good herd. The system is not an entirely new one, as it is practised to a great extent in some parts of Lincolnshire and other Northern counties, where there is not the excuse made for it in the States that it avoids the trouble and risk from the intense cold attending the farrowing of sows in the winter.

It may be that the severity of our winters is not usually great, but the cold, damp and foggy weather commonly experienced in England during the last two or three months of the year render it necessary to warmly house young pigs, and this is difficult in wooden houses of limited size, as these become hot and stuffy when entirely closed, or damp and cold when unclosed. Again, the labour attending the feeding of a large herd housed in isolated sties must be very considerable. Another objection raised against this farrowing of sows in these small houses is that it is difficult if not impossible at night to have the pigman in attendance on the sow, further, that it is not advisable to allow the young pigs to roam about with their dam until they are some weeks old, as when the weather is cold or wet they become chilled and when the sun is hot they quickly become blistered, both conditions materially interfering with their well doing.

It is claimed that both sow and pigs are able to secure a large portion of their living, but a sow with a good litter of pigs on her requires a considerable amount of food in addition to grass to enable her to do justice to her young, whilst the younger pigs are unable to digest any quantity of grass until they are some weeks old; besides this, the youngsters thrive much better during their early life when confined in quarters than when trailing about after the sow. Could we ensure fairly fine weather, and an absence of cold nights and very changeable weather, the little pigs' chances of thriving under outdoor conditions would be considerably enhanced.

Another alleged new discovery is the permitting of pigs to roam at large in woods and plantations, wooden huts or open sheds being provided as shelter. By this plan a considerable amount of pig food is obtained where the trees are not closely planted, so that grass grows freely, or, in the autumn, in the woods in which oaks, beech, hazel, or sweet chestnut form a portion of the trees. In such woods strong store pigs are able to obtain the major portion of their food, but where the trees are of a kind which does not produce nuts or are closely planted, the additional food must be more plentiful, whilst the manurial value of the food is wasted to a considerable extent.

Perhaps the most profitable form of outdoor pig-keeping is that of running the pigs in orchards. This system has many advantages, the pigs are able to live without much additional food for some months in the year, they consume the insect-affected fallen fruit, and so act as insecticides. The pigs also usually leave their droppings under the trees, which are thus benefited therefrom, and especially is this the case where the pigs are being fattened or fed on food which enables them to make flesh. Many years since, the writer had several customers for breeding pigs who kept numbers of pigs in their orchards. One fruit grower in Kent declared that fattening pigs in his orchard resulted in the growing of heavier crops of cherries of larger size, better colour, and finer flavour. Another whose apple orchard was disappointing followed my advice to fatten pigs in it, declared that the quantity of apples grown was much greater, whilst both the size and quality of these were infinitely better.

Under the modern system of pig-keeping it is more profitable to give some additional and concentrated food to the pigs having their liberty, it is therefore wise to secure the full benefit arising from the richer living by running the pigs where the manure can be utilised, and no better place than an orchard can be found, since shelter from sun and wind is furnished by the fruit trees, and the pigs deposit their urine and excrement in exactly the place where it is most urgently required.

The practice of growing considerable areas of rape or cole seed, artichokes, peas of various kinds, beans, etc., to be fed off by pigs is not followed extensively in this country, although pig-keepers in the United States, Canada, Germany, Denmark, etc., have a partiality to it, since it is declared to save labour and to bring the land into a good manurial condition for the growth of corn crops; still some few of our more advanced farmers have been in the habit of grazing off lucerne, clovers, and even permanent and temporary grasses by the aid of pigs, which have also received in addition a varying amount of roots, corn, or meal. It is asserted, and evidence is available to prove the truth of the statement, that land can be economically and quickly and vastly improved by following the system referred to above. The scarcity and high market value of miller's offals and of meals such as used in the past to be utilised to a great extent in the feeding of pigs, has caused pig-keepers to seek for other foods to take their place. The residuum from the crushing of palm nuts, cocoa-nuts, and ground nuts has been most successfully used in connection with various forms of vegetable food; even sows have reared good litters of pigs on about 2 lbs. of a mixture of the meals remaining from the extraction of the oil from the nuts mentioned, with the addition of some form of vegetable food. This last has comprised cooked potatoes, raw artichokes, mangolds, kohl rabi, swedes, cabbages, etc., during the winter months, and grass, lucerne, clover, vetches, cole seed, etc., during the summer months. Fattening pigs will require a somewhat larger quantity of concentrated food and a reduced amount of vegetable food. The pre-war belief that sharps or middlings only was the most suitable food for sows with litters and for newly weaned pigs has been somewhat modified. Whether or not the quality and price of middlings will be restored after the war and thus its use become general as of old, must be left, but it is probable that in the future a certain proportion of the meals referred to will continue to be used for both breeding and fattening pigs.

CHAPTER XV

PIG-FATTENING

If there be one task which is considered to be within the capacity of any individual, it is that of feeding a pig. In the good old times, the one thing needful was a good supply of barley meal, as much of this as the pig could possibly eat was placed into its trough each day until the pig was thought to be fat enough for slaughter. This was a very simple and at the same time a very costly process and was looked upon as the second of the two chief acts in the life of a pig. The first consisted of building up a frame on which fat could be stored. Just why these two processes were not combined has never been fully explained. One excuse made for this uneconomical process is that our forbears must have considered that there must be two distinct periods in the life of any animal intended for the food of man, that in which the structure was erected, and that in which the building was completely furnished with the material--flesh--in a state which most nearly satisfied the requirements or fancies of humanity. The system of first growing the frame and then packing it with flesh was not alone followed by the owners of pigs, as it was also adopted with cattle, which in the good old times passed three or four years in a state of semi-starvation ere they were placed on our best pastures to produce beef. Sheep, again, spent two or three years in building up their frames and in the production of a limited quantity of wool of inferior quality and strength, before they were considered in a fit state to make mutton economically. Another excuse which could have been offered by our forbears, but which is not now available, is that the cattle, sheep, and pigs of former times required age before it was possible to render them sufficiently fat for slaughter.

The very great improvement which has taken place during the past half century, in wellnigh every breed of pig, has deprived our present day pig-breeders of such an excuse, yet they persist in far too many instances in following the old-fashioned and uneconomical system of first growing the pig and then fatting it, whereas it is not only possible but infinitely more profitable to combine the two operations. So many persons have been in the habit of looking upon the pig as a mere scavenger or an animal to put out of sight certain articles containing a small amount of nutriment which, undisposed of, would become a nuisance or offensive to one or other of our organs. Even the pig itself has been considered by many farmers, especially those termed gentlemen farmers, as a necessary nuisance, whereas the pig is really a machine for the conversion of farm produce into meat, and like all machines, its output will depend entirely on the quantity and quality of the raw material, and the manner in which it is supplied. If the raw material be of inferior quality and supplied irregularly, or in too limited quantities, the article manufactured will be more costly and of an inferior quality. An extension in the time of manufacture means increased cost for fuel and for labour in attendance on the machinery. A certain quantity of fuel is being continually used in the furnace whether the engine is running at full power or at half power. It is exactly the same with the meat making machine, the pig every day of its existence consumes a certain quantity of food for which it gives one return only, its life. It has been conclusively proved that each pig weighing 100 lbs. requires 2 lbs. of food daily to enable it to sustain life, i.e. to replace loss of tissue, to provide heat, progression, etc., so that if a pig lives six months longer than is actually necessary to enable it to manufacture a certain weight of meat, it will have eaten to waste over 3 cwt. of good food.

A pig is like unto any other machine, it will produce the manufactured article most cheaply when it is fully supplied with the most suitable raw material. There is not the slightest doubt that the least costly pork is that which is produced by the pig which spends its whole time in the object of its existence, the manufacture of pork.

There is a further point of great importance. Wellnigh all those materials which are used in the feeding of pigs contain the constituents necessary for the building up of the frame and for the accumulation of fat or, as it is commonly termed, the making of meat. Evidently nature intended that the two operations should be carried on simultaneously. Those constituents which are required in the building up of the frame cannot be entirely used in the formation of fat, consequently if the frame is first built up and then an attempt is made to lay on flesh, a considerable portion of the building up constituents are simply wasted, since the pig has no need for them and cannot make complete use of them. They simply pass through the pig after taxing it to digest them, and are wasted.

Opinions and practices with regard to pig fatting have changed very much during the past half century, and especially so since the full effect of the fearful war has been felt. Rather before the first-mentioned period, the late Sir John Lawes, whose researches and experiments have been of lasting benefit to agriculturists, undertook to carry out experiments in connection with pig-breeding, and the result which appears to have impressed itself most upon the writers of the day was that barley meal was the best single food for the fatting of pigs. At the time named, our importations of maize and of many other materials now used in stock and especially pig-breeding were not of anything the magnitude of the period prior to the war, still, it seems to be strange to the enlightened pig-breeder of to-day that more serious endeavours should not have been made to determine the value of a mixed diet for pigs, since this had been proved to be beneficial and necessary in the case of human beings whose organs are so very similar to that of the despised pig.

Fortunately for us, and indeed for the stock-keepers in all parts of the world, experiments in the feeding of stock have been carried out in various countries, Denmark, Sweden, the United States, Canada, Germany, and indeed in nearly all countries, save to any great extent in England. In connection with pigs, the practices of a few of our more intelligent pig-keepers have been confirmed. Amongst these ideas which the old-fashioned ones looked upon as fads, was that of feeding pigs of all ages and especially fatting pigs on a certain proportion of vegetable food. Experiments have conclusively proved that the substitution of some 10 per cent of vegetable matter in place of an equal amount of meal or concentrated food, does not result in the slightest reduction in the live weight gain of the fatting pig, and further that the old idea that a limited quantity of vegetable food fed to a fatting pig tended to render the pork soft and to waste in the cooking was not founded on fact. Another fact which has evolved from these experiments is that the pig will make far greater progress on an equal amount of a mixture of foods than if fed solely on one food. This was clearly proved in many experiments as at the Wisconsin Agricultural Station, where one lot of pigs was fed on middlings alone, a second lot on corn meal alone, and a third lot on a mixture of corn meal and middlings. To make an increase of 100 lbs. in their live weight, the pigs in Lot 1 ate 522 lbs. of middlings, those in Lot 2 ate 537 lbs. of corn meal to make an equal increase in weight, whilst Lot 3, which were fed on a mixture of corn meal and middlings, required only 439 lbs., or a saving of one-fifth in the weight of food. In experiments with regard to the food value of corn meal and middlings carried out at the Missouri College, middlings also gave the best returns, but unfortunately the ages of the pigs used in the trials are omitted. This is important as middlings are considered to be of more value in the feeding of young than of older pigs, whilst the reverse holds good of corn or maize meal. Other trials were carried out at Wisconsin with the use of wheat meal alone as compared with a mixture of half wheat and half corn meal. In these the average quantity of wheat meal required for 100 lbs. increased live weight was 500 lbs., whilst only 485 lbs. of the mixture of wheat and corn meals was needed to obtain an equal increase or a saving of some 5 per cent was obtained by mixing the meals.

In the good old times it was considered to be the height of folly to make a change in the food on which the pigs were being fattened, yet our forbears would have been horrified had they been informed that it was imperative that they themselves should have no variety of food, that day after day the food at their various meals should be exactly similar; surely what is good for one animal should be good for another animal whose organs are of an exactly similar character. There is not the slightest doubt that advantage is derived from the variation in the food on which the pigs are being fattened. By this, it is not intended to suggest that a complete change of food should be made at stated times in the fatting pigs' food, as this would certainly result in a loss of time and food, but that a slight variation in the proportions of the different kinds of food is beneficial, or in the case where several different kinds of food are being fed as a mixture, another kind of food may be substituted so that the change made secures a variation which has the effect of whetting or enticing the appetite. A long continuance of the same kind of food has the effect of dulling the appetite. In addition to this, it is considered that a variation in the food tends to stimulate the digestive organs.

It is a mistake to allow too long a time to pass between feeding times; the pig is not endowed by nature with a capacious paunch which enables it to stow away a large quantity of food. Even the old system of feeding twice a day might be improved upon, and the fatting pig fed three times per day would make greater thrift, even should the actual daily quantity of food be not increased.

Again, so many persons are apt to give to the fatting pig a greater quantity of food than it requires or can eat with comfort to itself at one meal. Should this be pointed out to them, their usual reply would be that what the pig did not eat for their breakfast would be there in readiness for the evening meal unless they ate it during the day, as they frequently would do. This sounds plausible until the argument be closely examined. What would the pigman think if he were treated in a similar manner and an excessive quantity of food placed on his plate, and then at the next meals the stale food be again placed before him until it was finished? This certainly would not increase his appetite nor aid his digestion. Yet the most successful pigman is he who succeeds in so feeding his charges that they daily eat and thoroughly digest the greatest amount of food possible. In pig fattening, as in many other things, time is money. Further it is just as much a mistake for fatting pigs as for human beings to be continually eating, or at irregular intervals, small quantities of food. The two most certain indications that a lot of fatting pigs are thriving is to find that they are asleep and that their feeding troughs are empty. When pigs are fed a greater quantity of food than they can eat at once they will be frequently getting up to eat a little more of the surplus, and each time they rise from their bed they will evacuate their bowels, and in most cases before the major portion of the nutriment has been extracted.

Still another of the fallacies of our forbears was that the fatting pig made the greatest increase from a given quantity of food when it was at least approaching maturity and ripeness, or complete fatness. It was useless to argue with them, since anyone could see that it was so. If you suggested the use of the scales, the idea was scouted, since a person of any experience in pig fatting must be able to notice the increase in bulk of the pig. It is true that apparently the pig would be making a greater increase of weight as it approached the completion of its fatting process, since the addition to its weight and bulk would be almost entirely composed of fat which could only be deposited on the outside of the carcase. All the vacant space in the interior of the pig would have been occupied, the pig would have stored fat away in its muscles, around its kidneys, on its stomach, its bowels, and wherever it was possible to stow it away, but these additions to the weight of the carcase which had been proceeding in the early stages of the fatting could not be observed, nevertheless they were proceeding, and in this was the pig enabled in its early stage of fatting to make a profitable return for the food consumed.

Fortunately we are not left on this point to mere conjecture; many experiments have clearly proved that in the early stages of the life of a pig it is enabled to manufacture pork at a far less cost than in its later stages of life. The young pig also possesses over its older companion the great advantage of being able to eat and utilise a greater quantity of food in proportion to its weight or, in other words, the young pig can convert a greater quantity of raw material into the manufactured article than the more matured pig, in proportion to the amount of food required for the mere upkeep of the machinery. Experiments which most clearly prove this have been duplicated in Denmark, in the United States, etc. At Copenhagen nearly seventy different experiments were carried out with pigs of varying weights, with the result that pigs weighing about 275 lbs. live weight were found to require nearly twice as much food to make an increase in their live weight as did pigs weighing from 35 to 75 lbs. That this was not an exceptional case is clearly proved by the fact that the increase in the amount of food required to enable them to make an increase in their live weight was gradual, and shown in every stage; thus pigs of from 35 to 75 lbs. consumed 376 lbs. of food for each 100 lbs. increase; pigs of 75 to 115 lbs., 435 lbs.; pigs of 115 to 155 lbs., 466 lbs.; pigs of 155 lbs. to 195 lbs., 513 lbs.; pigs of 195 lbs. to 235 lbs., 540 lbs.; pigs of 235 lbs. to 275 lbs., 614 lbs.; and pigs of 275 lbs. to 315 lbs., 639 lbs.

Even if this series of experiments stood alone they surely would prove most conclusively that the common belief in old and nearly fat pigs giving the best return from the food consumed is founded on fiction, but similar tests were made at many of the American Experiment Stations, these tests together numbering some hundred. The results are given in tabulated form in Henry's Feeds and Feeding, where the various points are so clearly brought out that we have taken the liberty of lifting the whole of the notes relating to "weight, gain, and feed consumed" by pigs. "At many of our stations, records of weights and gains of pigs and feed consumed by them have been so reported as to permit of studies concerning the influence of increased size and weight of the animal on the consumption of food.

"All of the available data from trials of this character conducted in this country" (the United States) "up to the time of going to press, enter into the composition of the table given below. In compiling this table, six pounds of skim milk or twelve pounds of whey are calculated as equal to one pound of grain, according to the Danish valuation of these articles. For convenience of study, the data are presented for each period covering fifty pounds of growth, the actual average weight of the pigs, however, being given for each division:

DATA RELATIVE TO FEED, WEIGHT, AND GAIN OF PIGS-- MANY AMERICAN STATIONS

Weight of animals, lbs.	Average weight, lbs.	No. of stations reporting	No. of trials	No. of pigs in trials
15 to 50	38	9	41	174
50 " 100	78	13	100	417
100 " 150	128	13	119	495
150 " 200	174	11	107	489
200 " 250	226	12	72	300
250 " 300	271	8	46	223
300 " 350	320	3	19	105
350 " 400	378	1	5	36
400 " 450	429	1	5	36
450 " 500	471	1	2	18

Weight of animals, lbs.	Average feed eaten per day, lbs.	Feed eaten per 100 lbs. weight, lbs.	Average daily gain, lbs.	Feed for 100 lbs. gain, lbs.
15 to 50	2.23	5.95	.76	293
50 " 100	3.35	4.32	.83	400
100 " 150	4.79	3.75	1.10	437
150 " 200	5.91	3.43	1.24	482
200 " 250	6.57	2.91	1.33	498
250 " 300	7.40	2.74	1.46	511
300 " 350	7.50	2.35	1.40	535
350 " 400	8.52	2.25	1.98	431
400 " 450	8.18	1.91	1.71	479
450 " 500	10.00	2.12	1.77	562

"In the above table the large number of trials reported for pigs weighing up to 350 lbs. each furnishes reliable data. After this point is reached the number of animals is too small to give reliable averages. The heavy weight hogs reported in the last three lines of the table were fed by the writer (Professor Henry). They were mature specimens, with large frames and in lean flesh when feeding began, having been summered on pasture without grain. The figures are introduced to show what may be accomplished with mature hogs when they are in thin flesh at the beginning of fattening.

"We learn from the main portion of the table that from 105 to 435 pigs were employed in calculating each line of data. The number of trials furnishing the data varied from 19 to 119, and were conducted by from 3 to 13 experiment stations.

"Amount of food consumed daily by the pig. The sixth column of the table shows the average amount of feed consumed daily by pigs of different weights. From it we learn that pigs weighing less than 50 lbs. each, averaging 38 lbs., consumed on the average 2.23 lbs. of grain or grain equivalent, daily. As the animal increased in weight there was a gradual increase in the amount of food consumed, until we find the 450 lbs. hog eating 10 lbs. of grain daily, or more than four times as much as the 50 lbs. pig.

"Feed per 100 lbs. live weight: In the seventh column it is shown that pigs weighing 38 lbs. consumed 5.95 lbs. of feed for each 100 lbs. of live weight. This is about 6 per cent of their live weight. As the pigs grew larger they consumed less feed for 100 lbs. of live weight, until with the heaviest hogs the feed consumed was little more than 2 per cent of their live weight. Here was a decrease of about two-thirds in the feed consumption per 100 lbs. between early weight and maturity.

"Average daily gain: In the next column are presented data concerning the daily gain of the pig. It is shown that the 38 lb. pig gained .76 of a lb., or 2 per cent of its own weight daily. As it increased in size the pig made larger daily gains, the maximum being reached with those weighing 271 lbs., which made a daily gain of 1.46 lb. With large thin hogs the gain reached 1.98 lb., or practically 2 lbs. per day, but these animals, because of their mature frames and thin flesh, were fed under exceptional circumstances.

"Feed for 100 lbs. of gain: The last column is of interest to all, especially the practical feeder, for it teaches a most interesting and important lesson concerning the feed requirements of pigs. Those which average 38 lbs. each made 100 lbs. of gain from 293 lbs. of feed. This exceedingly small allowance of feed for gain was probably due in part to the fact that the young pigs used in these trials received much milk, which was practically all digestible, the other feed being also more highly digestible than that usually supplied older animals. With pigs weighing 78 lbs., 400 lbs. of feed were required for 100 lbs. of gain. There was a gradual increase of feed requirements for 100 lbs. of gain, until the hog weighing 320 lbs. required 525 lbs. for each 100 lbs. of gain. This is 135 lbs. or 33 per cent more feed than was required by the 78 lbs. pig."

These tables prove most conclusively that the idea which is almost universally prevalent that the fatting pig gives the greatest increase for the food which it consumes when it becomes matured and nearly fat is an entirely mistaken one, and that the young and growing pig, if well kept, not only eats more in proportion to its weight, but gives a better return for the food it consumes, besides requiring a smaller amount of food to keep life within itself, and to replace the certain loss sustained by movement, etc. There is still another point on which the young pig scores: its carcase realises a higher price per lb. on a majority of the markets. The fatting pig which pays best is one which has a short life and a merry one, never having to seek or wait for its food.

Amongst the many other questions which have been compelled attention owing to the shortage and the high value of pig food, is that of the advisability or the reverse of cooking the food given to pigs. When the practice of showing stock became fashionable every possible means of forcing the exhibits was practised, since early maturity was of so great importance, especially in the classes for the younger animals. The cooking of the stronger kinds of food such as old beans for horses had been found beneficial, as the risk of fever in the feet and other ailments had been greatly reduced by this practice. The stock man naturally concluded that the cooking or steaming of beans having proved to be of advantage, similar good results would follow the steaming of the other kinds of food. In this fanciful theory they would have been able to find ample support in many of the books on stock feeding which were published in the first half of the last century and even later. Like many other novelties, the steaming or boiling of almost all kinds of food for animals was followed in the establishments of well-to-do persons where cost was studied less than success in the show yards. Then, as now, the Germans took little for granted, they proceeded to test the much belauded new plan by attempting to discover the fact as to whether steaming rendered hay more digestible when fed to cattle, with the result that it was clearly proved that when the hay was fed dry 46 per cent of the protein was digested by the cattle while only 30 per cent was digested from the steamed hay. But as our present business is with pig-feeding, we will confine our remarks to the results of experiments carried out to test the effects of cooking the food of pigs. Perhaps the best summary of these is to be found in the most valuable work, Feeds and Feeding, by Professor Henry, who wrote Experiments with Cooked Feed for Pigs.

These have been so numerous that all cannot be here presented. Those given are selected because they are strictly representative, covering a wide range of country foods and conditions.

"At the Kansas Agricultural College, Shelton fed one lot of five pigs on cooked shelled corn, while a second lot of four, similar in all respects, was given uncooked shelled corn, the trial lasting ninety days. In cooking, the corn was placed in a barrel and water poured over it; into this mass a pipe carried steam, at a pressure ranging from 30 to 60 lbs. The kernels were cooked until they were sufficiently soft to be easily mashed between the thumb and finger.

"At the Iowa Agricultural College, Stalker conducted trials for 120 days in summer with cooked and uncooked shelled corn fed to Berkshire pigs.

"At the Dominion (Canada) Station, Robertson fed grade Chester Whites, a mixture of ground peas, barley, and rye, the trials beginning in December and lasting 141 days.

"At the Ohio Station, Devol fed pure bred Poland Chinas and Berkshires for 112 days in winter. One lot of three pigs received the meal cooked, while to the second lot it was given dry and uncooked.

"At the Wisconsin Station, the writer (Henry) has conducted many trials with cooked and uncooked feed for pigs. Only the later ones are here reported. These trials lasted from 56 to 84 days each, the kinds of feed experimented being given in the table.

"The five trials reported from the Wisconsin Station, as will be seen by consulting the table, are slightly in favour of cooked food, the difference being very small, however. These are the only feeding trials reported from any experiment station, so far as known to the writer, where the results are favourable to cooking. Ten other trials by the writer with cooked and uncooked feed for swine all gave results unfavourable to cooking these, and a number of trials at other stations with cooked and uncooked feed for swine are not included for want of space."

A table showing the stations at which the various experiments were carried out, the numbers and weights of the pigs, the varieties of foods, the duration of the different trials, the daily gain, the weights of cooked and uncooked food consumed, the manner of cooking, the total increases in weight and the quantities of cooked and uncooked food required for increases of 100 lbs. in the live weights of the pigs are given. Professor Henry sums these up and writes: "Including all the trials then, so far as is known, that have been favourable to cooking feed and omitting many for lack of space, that are unfavourable to that operation, the average shows that 476 lbs. of uncooked meal or grain were required for 100 lbs. of gain with pigs, while after it was cooked 505 lbs. were required. This shows a loss of 6 per cent of the feeding value of these substances through cooking."

Some thirty-five years since the present writer made some small experiments in the feeding of cooked and uncooked whole maize; in each case it was found that the pigs ate a greater quantity of uncooked than cooked maize, and made a greater proportionate increase in weight from the food consumed. Only one opinion appears to be possible, and this is that the cooking of food for pigs, save potatoes, entails a loss of time, an increase in cost, and a reduced return.

CHAPTER XVI

A PIG CALENDAR

The pig-keeper, like the gardener, seldom has to seek for employment, indeed his work may be said to be only occasionally completed. There are always many little odd jobs to do, which if neglected may result in loss, or a greatly increased amount of work at some later period. The old proverb "A stitch in time saves nine" is equally as true in connection with pig keeping as with any other form of work.

In years gone by the month of January was considered to be quite a slack time for pig-keepers, the sows and the store pigs usually found the greater part of their living in the yards where the cattle were fed on the straw which was continually being placed in the cribs as the old-man-of-the-farm threshed the corn out of it with his flail. Many of the cribs had slatted bottoms so that any kernels of corn which were left in the straw would drop through and be picked up by the pigs which found their way under the cribs. In most of the old-fashioned large yards a corner would be railed off in which the pigs would be given a few turnips, swedes, or small potatoes, and occasionally a handful or two of beans or even a sheaf of beans. Those fatting pigs which had not already been converted into bacon for consumption in the farm-house were fed mainly on meal ground at the local wind or water mill from the tail corn grown by the farmer. At the present time the most up-to-date pig-keepers so arrange that many of the older sows farrow during this month of January so that the sows have their second litter of the year late in the month of June or early in July in order that both litters of pigs obtain the greatest amount of benefit from the growing and hot season, since pigs thrive best when the days are lengthening and when the sun shines.

Of late years we appear to have had somewhat severe weather in January. This has rendered it the more necessary that care should be taken in providing water and wind-tight sties, in which the sows farrow. Warmth with free ventilation is needed. The latter is particularly necessary after the pigs are a few days old, as these do not suffer so much from cold as they do from damp and draughts. Of course whilst the sow is farrowing warmth is imperative, as the moist little pigs when first ejected very quickly become chilled in severe frost, unless they are promptly wiped with a dry cloth, allowed a draught or two of new milk from the sow, and then placed in a box or hamper three parts filled with dry wheat straw. When once the pigs become thoroughly dry the cold does not affect them very much, providing that the sow furnishes her family with a full supply of milk. The cost of heating a little water so that the sow and also the young pigs as soon as they begin to eat may have warmed food, will be slight, as there is nearly always a fire required in cottage and farm-house during the cold weather. Warm food makes a vast difference in the thrift of pigs, especially of young ones. Very slight observation will reveal the marked difference in the comfort of a pig which has had a meal of warm or of cold food. In the former case the pig will return to its nest and is soon lost in sleep, whilst the poor beggar which has had its breakfast on cold and occasionally frozen food will be the picture of misery and shaking with cold, much of its natural heat produced from its last meal being required to warm up the food ere its digestive organs can commence work. Coal and wood are at all times less expensive to warm up food than the animal fat which is burned in nature's lamp.

Provision should have been made for the supply of some kind of vegetable food which pigs require, particularly when in confinement. Kohl rabi, swedes and cabbages, of which the first named is the best, are all suitable, but the most nourishing are artichokes, which like the three former should be fed raw, and potatoes which should be cooked ere they are fed to the pigs. The difference in the feeding value between cooked and uncooked potatoes is great. It is scarcely necessary to point out that all vegetable food fed to pigs should have been protected from frost.

The operations connected with pig-keeping are very similar in February to those of the preceding month. Towards the end of the month kohl rabis will have lost much of their feeding value. On sunny days a run out for a few minutes will be of great benefit to the young pigs over a month old; as soon as they cease to gallop about they should be shut up again, as if allowed to lie down they may contract a chill which might result in "cramp" or rheumatism. Sows with litters two or three weeks old should be allowed out of the sty each morning and afternoon for a short time.

The month of March brings with it an extra amount of work for the pig-keeper, who will now think of selling the pigs born early in January unless he purposes to keep them on and have them ready for sale as fat pigs in harvest time, when there is always a good demand for medium sized fat pigs. Anyway the sow pigs intended for breeding will have been picked out and earmarked, this last should not be neglected after the others have been spayed.

This last operation has of late years been much neglected; this is a great mistake, as experiments have clearly proved that on an average sow pigs which have been spayed will make an equal gain in live weight on 5 per cent less food than will an unspayed sow pig, when both have become some five or six months old, and the periods of [oe]strum have commenced.

The sows which farrowed in January should now be weaned from their pigs, and should be ready to be mated within a few days. The sows should be carefully watched for the signs of heat or restlessness. Some sows give little indication of this unrest, which is almost certain to appear within four or five days providing the sow is in a healthy and vigorous condition. To miss the sow means a loss of three weeks of most valuable time, besides the risk of trouble in getting the sow to conceive after she had been baulked. With the passing of the month swedes and artichokes will have lost much of their nourishment; mangolds can now take their place. It is a good plan to expose the mangolds to the air for a few days prior to feeding them to the pigs; this exposure hastens their ripening and reduces the proportion of water. Of course care must be taken to prevent them becoming frozen, as in March this might be the case.

In the Southern counties tares, lucerne, and grass are sufficiently forward towards the end of April to be cut and fed to the pigs which are confined in the buildings. The pigs both fat and store will fully repay the cost of labour in the cutting and carting of these vegetable foods. Brood sows both in pig and with litters dependent on them, should be allowed their liberty in the grass fields. This will both greatly reduce the cost of keep and tend to their thrift and well doing. Young pigs over a month old should have a run out both morning and afternoon. Newly weaned pigs which have been well done are always in keen request in the months of April and May at prices higher than in any other portion of the year, owing to the demand from the cheese-makers who have a superabundance of whey, of which 12 lbs. when fed in proper combination is considered to be equal in value to 1 lb. of meal. Unfortunately, so many dairymen do not study the requirements of the pig, and imagine that it will give a good return from an excess of liquid in the form of whey. Without some concentrated food the pig will not thrive on whey. Numbers of young pigs are also required in those districts where butter-making is carried on to consume the butter milk, and in ordinary times much of the separated or skim milk. In the feeding of this again the results are not so good as they should be owing to neglect. Both foods have been rendered unbalanced owing to the extraction of the butter fat, so that although new milk may be fed alone, the others require additional food which should contain some oil or fat to be fed with them, or they cause indigestion and want of thrift, particularly in young and immature pigs.

The roots of all kinds, save potatoes and mangolds, have ceased to be of much value before April ends, vetches and lucerne will prove to be the best of substitutes. Spring cabbages are generally of more value for human consumption than can be obtained from their use as pig food. If there be any grass land available the in-pig sows and the stores, should there be such, should now find the major portion of their food out of doors.

As a rule far too little attention is paid to the growth of lucerne in this country. It is undoubtedly one of the most nutritive of our vegetable crops. It also produces a large weight of food extending over several months, and continues fruitful for many years providing attention is paid to the keeping it free from grasses. It has the additional advantage of furnishing a full supply of food when the weather is so dry that grass and some other foods produce little. It is true that in the initial stage it requires time and care, but the results from it amply repay both. One of the best seasons for sowing it is the month of May. The operation is simple, the land having been cleared the seed is sown in drills about 1 ft. apart, the quantity of seed required being at the rate of 20 lbs. per acre, say 2 oz. per pole or a drill 35 yards long. As soon as the plants are high enough the land should be hand hoed, and if kept free of weeds a light crop can generally be cut from it towards the end of August. In the following years it will produce at least three cuttings annually.

Some persons are of opinion that as lucerne is such a deep-rooted plant manure is unnecessary. It is true that the roots penetrate several feet into the soil, still an application of short manure or rotted vegetable matter applied each autumn will give a good return.

The chief point in the use of lucerne for pigs and in the production of a maximum crop is to cut it when young. The pigs will thrive on it far better in this state than when the stalks become hard and sticky. In the latter stage it is likely to cause constipation. It is best not to graze it with either horses, cattle, or pigs, but benefit to it results from folding it in the autumn with ewes or other sheep which find most of their other food on the stubbles, commons, heaths, etc.

All the sows and the yelts intended for breeding should now spend their whole time out of doors. It might be noted that lucerne will grow on almost any kind of land providing it is well drained--stagnant water destroys it.

The duties of the pig-keeper are very similar in the month of June to those of the previous month. Prior to the outbreak of war it was becoming general amongst the most practical pigmen to continue to fatten pigs all the year round. The old-fashioned idea that pork was not a suitable food during any of the months in which there was not the letter "r" had become exploded. Not only did the bacon curers require a supply of fat pigs weighing from 200 to 220 lbs. alive, but there was a good demand from the butchers for small fat pigs weighing from 80 to 140 lbs. alive.

It must not be forgotten that a smaller quantity of food is required to produce a pound of pork during the summer than during the winter months. This has been clearly proved in many experiments. The difference varies according to the temperature. In the very cold weather experienced in some portions of the United States it was found that some pigs actually made no increase in weight when well fed, the whole of the nutriment having to be utilised in keeping up the bodily warmth of the pigs.

The months of July and August see little change in the duties of the attendant on pigs. The old-fashioned plan of running the pigs on the corn stubbles has almost gone out of fashion. The improved system of harvesting the crops leaves less corn on the land, whilst the cost of labour in keeping the pigs is almost prohibitive. At one time there used to be a keen demand for young pigs in the month of August for so-called "shacking" or running on the stubbles. Experience has proved that these pigs pay less frequently under present conditions than they did under the old ones.

The scarcity of vegetable food which usually shows itself in August is now, in September, met to a considerable extent by the plan of the early digging of potatoes. Large quantities of chats, and sometimes of slightly diseased ones are now cooked and fed to the pigs with a certain proportion of meal. As a rule there is a keen demand for pork in the month of September. Towards the end of the month all pigs should be under shelter at night.

During the last three months of the year there is little variation in the management of pigs. One of the common mistakes made by farmers is to neglect their pigs in the autumn, at the very season when a little extra food is needed, and for which the pigs will give a better return than at almost any time of the year. The early portion of October is one of the best periods for mating the sows, the yelts may be left until the latter part of the month so that their pigs do not arrive until the month of February when the days are lengthening and the sun has more power. It is advisable to have many of the fat pigs ready for market ere the month of November ends, as the demand for pork is usually slack for two or three weeks prior to and after Christmas.

CHAPTER XVII

DISEASES OF THE PIG

Fortunately, the pig is subject to comparatively few serious diseases--save swine fever, swine erysipelas, and very occasionally anthrax, which are contagious or infectious, and all in the special charts of the veterinary department of the Board of Agriculture, and within the contagious Diseases Animals Acts. Prior to the stamping out of Foot and Mouth Disease or apthous fever and rabies, pigs suffered from these contagious and infectious diseases, particularly the former of the two, which caused immense losses, especially of young pigs, during the latter half of the past century.

Of the other ailments to which pig flesh is heir, the majority and the chief of them are mainly due to that want of knowledge or care in the feeding and in the housing of the pigs which renders them more susceptible to the sudden changes in the temperature or to the inclemency of the season. In former chapters some, if not all, of these ailments have been referred to, but it may be more convenient to our readers to include in one chapter a brief description of the ailments and the remedies and means of prevention.

SWINE FEVER

Some thirty years since the losses from this disease were of so serious a nature that the Board of Agriculture determined to attempt to stamp it out, as they had succeeded in stamping out pleura pneumonia in cattle, and foot and mouth disease. The success of their efforts was not at all commensurate with the outlay. The failure was attributed to many causes; amongst them the want of a complete knowledge of the disease, the impossibility of diagnosing it during the life of the patient, the absence of sympathy on the part of the local veterinary surgeons owing to certain steps taken by the then Veterinary Adviser of the Board, to which further reference is now inadvisable, and to the general opposition of pig-keepers who had as little faith in many of the post mortems and their results as in the power of the authorities to stamp out the disease which under various names had been more or less common in the country so long as they could remember. Doubts were also passed on the infectivity or contagiousness of swine fever, or as it was variously termed red soldier, spots, etc.

This disbelief was probably due in part to the fact that some of the external symptoms of swine fever, swine erysipelas, and heart disease, such as discoloration of the skin were of a similar character. In some instances this redness of the skin, which was looked upon as a sure sign that the pig had died from swine fever, did not prove to be infectious, as no other cases followed amongst the in-contact pigs. This led to the general belief that swine fever was not necessarily infectious. Dissatisfaction with the arbitrary manner in which the restrictions in movement, etc., were carried out did not mend matters, nor help to render the efforts of the Board more successful.

At the present time it is imperative on the part of the owner of an ill pig to report the fact to the nearest policeman. The owner then merely carries out the instructions supplied to him by the police so that it is almost unnecessary to state that the symptoms of swine fever are several. At times the attack is of so virulent a nature that a pig may take its food all right in the afternoon and be dead the next morning, no discoloration of the skin or other external symptoms being visible before or immediately after death.

As a rule when the pig is attacked the first symptom is loss of appetite, generally accompanied by a feverish condition of the skin which shows more or fewer red spots behind the shoulder, and inside the thighs, or in those portions of the body where the skin is the thinnest and most free from hair. The great desire of the affected pig is to burrow into the litter and to remain undisturbed, save when the feverish thirst impels it to seek moisture of any sort or kind, even urine which may have settled into any unevenness of the floor of the sty.

Many of the ailing pigs suffer from a dry, husky cough, a gummy discharge exudes from the eyes and forms a ring round them, the ankles become affected, and the muscles of the back become weakened so that the pig has difficulty in walking. The discoloration of the skin may or may not increase, but the weakness gradually becomes greater so that death may follow within a day or two from the first attack. Occasionally the affected pig will continue to live for several days, and eventually recover so much that it can be fatted, but there exists a great risk of the recovered pig being what is termed a "carrier" of the disease, and possessing the ability to infect other pigs with which it may come in close contact, although the germs of the disease which it carries do not affect its own health. Similar instances of human beings being "carriers" of the disease have been recorded. So difficult is it at times to discover the source of the infection of swine fever that certain persons who are not amongst the strongest believers in the practical knowledge of the members of the veterinary profession assert that swine fever need not necessarily be the result of infection, but that injudicious feeding or the neglect of sanitary arrangements will sometimes cause an outbreak. There does not appear to be the slightest ground for this belief, as there is a specific virus which when it obtains ingress into the body of the pig, whether by the mouth, nose, or in any other way, may result in an attack, more or less severe, of swine fever, unless the virus has become so attenuated that it is unable to affect the host sufficiently. This attenuation, which is due to causes which are probably not completely known, is commonly the cause of the absence of further cases of swine fever amongst one of a lot of pigs which has had a very mild attack. This variation in the virulence of most infectious diseases has been noticed and recorded.

At the present time the Board of Agriculture have suspended the slaughter order in cases where the owner of the pigs desires to inoculate the in-contact pigs with serum which is supplied from the Veterinary College. The experiment has not been in operation sufficiently long enough to express a confident opinion upon its results, but it is stated that in Denmark the inoculation of the pigs which have been in contact with diseased pigs has proved to be a success. The risks of carrying out the experiment are by no means slight, but appear to be worth running if there be any great probability of success.

SWINE ERYSIPELAS

The symptoms of this disease, which fortunately is not so common as swine fever, owing probably to its being more fatal and in a shorter time, are very similar to those of swine fever, save that the husky cough and the weakness of the muscles of the back are generally absent. The post mortem shows distinctive differences from those of swine fever. There appears to be far greater difficulty in thoroughly disinfecting the sty in which pigs suffering from erysipelas have been housed than after swine fever cases; not only so, but the virus remains active for a very long period, so that any accident which may expose the virus even after many months may affect any pigs with which it comes in contact.

In an outbreak of swine erysipelas it is advisable to have the unaffected pigs inoculated as well as those housed in a sty or building in which at any time pigs suffering from erysipelas have been housed. A certain limited number may die, and a few suffer for a time, but the total loss will be considerably reduced.

ANTHRAX, FOOT AND MOUTH DISEASE AND RABIES

It may be unnecessary to describe these very infectious or contagious diseases to which pigs are subject, as fortunately the steps taken to stamp them out, and which were much decried when taken by the Board of Agriculture, have proved so successful that the two latter are stamped out, and the first named is so promptly and effectually dealt with that a case of it amongst swine is seldom recorded.

CRAMP, DIARRH[OE]A AND EPILEPTIC FITS

These diseases, which are more frequent amongst young pigs, have been fully described in the chapters dealing with the rearing, weaning, and growing of pigs, where it is pointed out that they are all mainly due to faults in feeding, and the simple remedies applicable are there given.

Hernia and Scrotal Hernia are also treated upon in the chapter on the Farrowing Sow.

INVERSION OF THE VAGINA OR THE UTERUS

These two troubles, of which the latter is a complete expulsion and the former only a partial protrusion of the "breeding bag," are generally the result of a difficult or a protracted farrowing. The second is almost impossible of treatment, and indeed may be declared as fatal, so that the loss may be reduced by prompt slaughter.

The first varies in extent; a partial or limited inversion may at times be noticeable during the latter stages of pregnancy, and then after delivery may disappear without treatment until the pressure due to the increasing size of the f[oe]tus again causes it. Even in serious cases which attend the delivery and are due to excessive straining of the sow, the attack is not necessarily fatal if extreme care in treatment is applied. The first thing is to wash the protruding part with warm water, to which some disinfectant has been added, in order that all dirt, short straw, etc., shall be removed. The sow should then be made to rise, or if she refuses, as is not uncommon, the hind quarters of the sow should be raised and the protruding portion be gently but firmly forced back. In order to prevent a re-expulsion stitches with strong cord or leather lace should be inserted into the edges of the vulva--these need not be very close together or otherwise the sow would be unable to make water. For a few days the sow must be kept as quiet as possible and fed on a little nourishing but laxative food, so that the pressure on the vagina is slight until the muscles regain their normal strength. Should there be the slightest symptom of constipation, salts or castor oil should be given to the sow. No harm, but rather good, will attend the giving of a gentle dose of salts at the first time of feeding after the operation as there is certain to be an amount of inflammation present.

INVERSION OF THE RECTUM

This expulsion of the gut as it is commonly termed is not often experienced amongst mature pigs. Young pigs are not uncommonly affected save when constipation is neglected, or when the food is of a heating nature which causes continual difficulty on the part of the pig in expelling the f鎔es. The effort of straining causes the gut to exude. Similar treatment, save as to the stitching of the part, as with inversion of the vagina, should be followed.

TENDER FEET

This trouble is frequently mistaken for cramp or rheumatism, and is generally due to the same causes, injudicious feeding, etc. In the latter disease the ankles are mainly affected, in the case of fever in the feet, the feet only are affected. A strong dose of Epsom salts should be given and daily doses of nitre should be given in the food. The object should be to reduce and remove the fever and then to cure or remove that tenderness and soreness of the feet which follows the fever. Poulticing the feet and applying diluted white oils by adding equal quantities of water and vinegar around the coronets are both remedial measures of great value.

CONSTIPATION

This trouble is very common amongst pigs which are confined to the sties, its avoidance is comparatively easy, when the want of exercise is the sole cause. A run in an enclosure or even in the road will almost always result in the pig evacuating dung and water. A dose of salts, varying from 1/2 oz. to 1-1/2 oz. for each pig, according to age, in the next supply of food is advisable.

Constipation is usually the first indication of many of the troubles to which the pig is heir. The little pig on its mother becomes constipated when the food fed to the mother is unsuitable, and the pig suffers from indigestion; fever caused by a chill is also foretold by constipation which should be first removed by a gentle dose of salts or of castor oil; the last only to be used in severe cases. Linseed oil is also frequently used to relieve the constipation, but with this there is a fear of billiousness following its use. If exercise and the above remedies do not effect a cure, an enema of soap and water or even glycerine may be necessary. Old-fashioned pigmen remove the hard and knotty f鎾 es by the aid of the finger.

ECZEMA

This is sometimes called a skin disease, but it appears to be rather a symptom of a severe attack of indigestion or of billiousness than a disease in itself. It shows itself in the form of a bright red spot, varying in size from that of a threepenny piece to that of a shilling, these spots vary greatly in number. Small pimples appear on the spots from which a sticky fluid exudes. As soon as the bowels are thoroughly relieved by aperient medicine, the spots become dark in colour and peel off the skin. The application of oil to the spots hastens the shedding of them. A dose of sulphur of one to eight drachms in addition to the salts will be beneficial.

Frequently the pig will refuse to eat, it will then be necessary to dose it. The pig must be caught, its head raised and the liquid gently poured down its throat, the greatest care being taken not to pour the liquid whilst the pig is squealing or the medicine will go into the lungs and cause suffocation, or inflammation of the lungs which will generally prove fatal.

MEASLES

This is a trouble of a very similar character to eczema save that the red spots are more numerous and of a more irritating character. The patient is continually rubbing itself against the wall or any prominence in an endeavour to relieve the itching. The pig is also more feverish. The pig should be placed in a warm sty, with plenty of dry straw, into which it will quickly burrow. A dose of Epsom salts to which is added a small quantity of spirit of nitre should be given, as the pig affected will almost invariably refuse food for a time. Neat's foot or sweet oil applied to the spots will relieve the irritation.

RICKETS

This is not by any means a common ailment amongst pigs, but it is very hereditary. The most common cause is too close breeding. The bones and joints appear to be unequal to the performance of their duties, the pig staggers and stumbles when it attempts to move, whilst sometimes the back is affected, when the pig is stated to be suffering from "swayback." As a rule treatment is inadvisable as recovery is doubtful. The first loss by knocking the pig on the head is generally the least.

TUBERCULOSIS

Pigs, like unto human beings, are much subject to tuberculosis when they are kept under conditions similar to those which result in human beings becoming affected. The disease is highly infectious, pigs coming in contact with or even being housed in sties where pigs affected have been recently kept are very likely to become infected. Some persons declare that tuberculosis, or, as it is more commonly called, consumption, is hereditary. For this there does not appear to be any foundation. The chief thing to prevent one's animals being affected is to keep them away from contagion. Although many parts of the body may be attacked by tuberculosis, the lungs are more frequently affected than any other of the organs, owing probably to the ease with which infection by the minute germ is conveyed to the lungs in the act of breathing.

In the past a considerable number of pigs became infected through being fed on skim milk which contained germs from the udder of a cow suffering from a tuberculous udder. In these cases of the lungs and the bowels becoming tubercular, the pigs become unthrifty and frequently waste away and die. When the bones and other portions of the body are attacked the development of the disease is not so rapid, but in any case the wisest plan is to destroy the animal and thoroughly disinfect the place in which it has been kept. Save when the disease is local and of very limited duration the meat of a pig suffering from tuberculosis is unfit for human consumption.

WORMS

Pigs are subject to various kinds of worms. Of these the most serious by far is the worm which causes the disease called Trichinosis in man. The worms are transmitted to man in pork from a diseased pig. Thorough cooking of meat appears to destroy the vitality of the worm, but in foreign countries where the pork is eaten in an uncooked or an undercooked condition the disease is not uncommon. Fortunately, Trichinosis is almost unknown in this country, owing to our more stringent sanitary conditions, the disease being due in the pig to the eating of human excrement in which are thread worms.

The most common kind of pig worm in this country is the round white worm, pointed at both ends. Its length varies from one to several inches. Its presence is often unsuspected until one or more of the worms are noticed in the dung of the pig. It is readily got rid of by keeping the pigs from food for at least twelve hours, and then giving them a little tempting food in which a dose of santonine, varying from three to ten grains for each pig, according to its age, has been added. Some two hours later a dose of castor oil of from 1/4 oz. to 2 oz., or of one to two ounces of Epsom salts, should be given in milk or some other tempting food. Similar treatment will prove successful in the case of pigs affected with the smaller kind of worms save that of the worm which causes what is commonly known as "husk." This worm makes its home in the windpipe and bronchial tubes. It is advisable to obtain from a chemist a drench for the riddance of this worm, as the remedies will consist of linseed oil, turpentine, spirits of camphor, and asaf[oe]tida.

SORE TEATS

Occasionally the teats of sows, especially sows with their first litters, become chapped or sore. This trouble is frequently due to the too vigorous sucking of the little pigs when the supply of milk is short, to the biting of the teats when the sharp little teeth have not been broken off, or even to cold winds.

An application of boro-glyceride will usually effect a speedy cure. In persistent cases it will be advisable to give the sow a dose or two of opening medicine such as salts or sulphur.

SALT AND SODA POISONING

Although these can scarcely be classed as diseases, the effects are often more serious than those of some actual diseases to which swine are more or less subject.

In the majority of cases the cause is the neglect of the cook to keep separate from the swill the water in which salted meat or other food has been boiled, or the water to which soda has been added in the washing of the plates, etc. An attack if at all severe is usually fatal.

The symptoms are a discoloration of the skin, and a refusal of food. As these are the usual symptoms of several other ailments, it is difficult to determine the cause of death save by a post-mortem examination. It is to be feared that this mixing of a solution of salt and soda with the other swill will be one of the difficulties met with in the more general utilisation of kitchen refuse in the keeping of pigs.

CHAPTER XVIII

THE CURING OF PORK

In the good old times bacon curing was carried on in the large majority of farm-houses as well as in many houses in the country districts, not only where there were conveniences for the keeping of pigs, but many householders were in the habit of buying carcases of pork from their neighbours and curing the major portion for the following year's supply of cured meats. Even the better class labourers would kill and cure it so that as long as it lasted they had on hand a supply of most nutritious and suitable food. Unfortunately a great change has taken place of late years; this convenient and profitable plan has been superseded. The causes may have been many; amongst them, the importation of immense quantities of salt pork of very inferior quality at very low prices from the United States; the change in the public taste which is now for mild cured and lean bacon from young pigs, instead of the more heavily salted meats from older and fatter pigs; the great decrease in the number of pigs kept by cottagers and others in urban districts through the operation of the so-called sanitary regulations; and probably from the different style of living, which may or may not be an improvement, amongst the residents in country districts.

It may be that one of the many changes which have been brought about by, and which will also follow, the war will be a return to the more simple and less luxurious manner of living. It is certain that a more economical system will have to be followed, and one of the means of effecting this may be a return to the keeping of pigs during their growing stage on the house and garden refuse, and then when the pigs have been fattened, by the killing and curing of the carcase for home consumption.

Much has been written during recent years about the folly of allowing so many millions of sovereigns to go out of the country in payment for the vast weight of bacon, hams, and lard which we import from foreign countries. Residents in the country have been blamed by town residents and literary men for their alleged want of enterprise in not breeding and fattening the few extra million pigs which would furnish an amount of pig produce equal to that imported, and thus, as they declare, save the country that outlay which is a dead loss to these islands.

It may at once be frankly admitted that a very considerable increase in the number of our pig population is possible without any very greatly extended cost of food, but when it is contended that farmers and even cottagers are grossly neglectful in not producing sufficient pork and its products for the use of the whole of the population of these islands, an injustice is done, as the breeding and feeding of pigs is a business calling, not a philanthropical pursuit. Farmers and cottagers are like other manufacturers of necessary articles; they produce in order to live, and they cease to manufacture an article when its production ceases to repay them for their outlay and trouble. They must of necessity do so, or they come to grief and are unable to carry on their farms or businesses.

It matters not what the cause be for the ability of the foreigner to produce and land on our markets articles cheaper than we can afford to offer them at, the result is the same--the home production is automatically reduced. There are many causes which have helped to render it possible for foreigners to supply us with a certain proportion of the pork and bacon which we require at a less cost than our home breeder and feeders of pigs can supply it. These include help to the farmers from the Governments of certain countries such as Denmark, where assistance is given in the purchase of pure bred pigs for the improvement of the native pigs, in the reduced railway and other rates on the transit of pigs, foods, and bacon, in the provision of certain foods, and in carrying out experiments in order to show how they may be utilised in the best manner. Stud farms have also been established from which pure bred boars are distributed, whilst the whole industry of pig breeding and bacon curing is carried on under the supervision and with the advice of many Government officials appointed for the purpose. The intrinsic value of this assistance is perceptible, as in no other country are pig-keeping and bacon curing carried on with greater monetary success than in Denmark.

It is also asserted that the general system of farming in Denmark has also contributed very largely to the phenomenal prosperity of the pig industry, in that a very large proportion of the land is owned and farmed by comparatively small farmers, men who have a direct interest in the improvement of the land, and who with their families perform the major portion of the work on the land and in attendance on the stock. The land is almost certain to be well managed and the stock to receive the best possible attention with, comparatively speaking, little cost as to labour. The animals on the farm are likely to be of a higher grade and the returns from them of an increased character, than when strangers and disinterested hired labour attend and feeds them.

Another the great advantages possessed by some of our foreign competitors is the very much better supply of feeding stuffs and their very considerably lower cost. Take the United States, for instance, the enormous supply of maize alone enables American pigmen to manufacture pork at a cost which enables the packers to land bacon, hams, and lard on the British shores which our home pig producers cannot approach. Although it cannot be said that the cost of labour is less in the States than in England, yet there are some countries from which we import pork products where the labour is far more plentiful and less costly. In the future the allowance for labour will have to be on a more liberal scale than hitherto when estimating the cost of producing pork, unless the number of persons owning and occupying small holdings is greatly increased.

It has been stated that our home producers of pork and bacon will obtain a considerable advantage in the future in that the freight on the imported meats will be so much higher. It is most probable that this will increase the expense of landing bacon, etc., on our markets; on the other hand, as we import so large a proportion of the pig fattening foods, the cost of food will most likely be increased to quite the same if not to a greater extent. The only plan to reduce this extra expense will be to lessen the outlay on imported foods by paying more attention to the growth of various foods suitable for pigs, attending more carefully to our pigs and feeding them on common-sense lines. In these particulars there is room for much improvement in many piggeries.

By reducing the cost of the production of pork and by the more general adoption of the system of home curing we shall not only obtain our bacon at less cost, but we shall have a far greater amount of the finest quality of bacon and hams generally available. We imagine that the reader of the earlier portion of this book will experience little difficulty in producing fine quality pork at a minimum cost--it will then remain to cure and dry it properly.

The fattened pig should not be fed for some twenty-four hours before it is killed; after slaughter the carcase should remain hanging until it is thoroughly cooled. The manner of cutting up will depend on the custom in the particular district. In some parts of the country the pig is split down, the head, feet, and tail taken off, the leaf and kidneys and the skirt taken out, the loin and the crop with a certain proportion of the lean cut off, and in some cases the shoulder blade is drawn; after the necessary trimming a Wiltshire side remains.

In other districts the ham and the shoulder are cut off and the side is converted into a middle, a ham and a shoulder or fore-ham. The jowls are taken off the head and salted with the bacon and hams. The upper part of the head, or, as it is commonly termed, the scorf, is usually used with the feet in the manufacture of brawn, or, as it is sometimes called, pork cheese--presumably from its being cooled in a form, and then turned out on to the dish on which it is served at table.

The first operation in curing is to distribute a small quantity of salt all over the meat to be cured. If allowed to remain about forty-eight hours the blood remaining in the meat will have become dissolved, and will have exuded from the carcase. This liquid should be thrown away. A mixture in the proportion of 4 lbs. salt, 1 lb. coarse brown sugar, 1 oz. saltpetre, 1/4 oz. bay salt, and 1/4 oz. salt prunell should be prepared, and a portion of it be applied to all parts of the meat and particularly in the pocket hole, if the shoulder blade has been drawn. This should be continued for from twenty to thirty days, according to the thickness of the meat and the degree of saltness desired. In one or two districts of a limited area it is usual to rub the meat somewhat violently with a large pebble when applying the salt mixture, the alleged object being to rub in the salt; but for this there is not the slightest necessity as the result of the rubbing is nil, since the salt will penetrate the meat equally as well without the manipulation as with it. The principal point is to secure the distribution of the salt to every part of the meat so that the salt can penetrate and preserve it.

When sufficiently cured the meat should be hung up and dried. If it be desired to have it smoked this is best done at the village bakery or smoke drying house. Smoking of hams and bacon is possible on a small scale with the aid of a smoke oven such as supplied by Messrs. Douglas and Sons of Putney, but it is, as a rule, cheaper and less troublesome to send the meat to the village smoking house. It will be advisable to brand or otherwise mark each piece of cured meat sent to be smoked, as the return of the same pieces is thus assured.

Where the home curing of bacon and hams is followed, this is best carried out from the middle of October to the end of March; if it be attempted earlier or later a cold chamber is necessary.

The manufacture of salt pork is carried on all the year through as the meat is usually kept in the brine, where it will keep perfectly good for a considerable time providing it is perfectly sweet when first placed in the brine. To secure this it is advisable to have the pig killed in the evening, covered over with a cloth to prevent the flies approaching it, and hanging it in a cool place so that all the natural heat has escaped ere it is cut up and placed in the pickle pot. It may be advisable to note that the last is only possible with a small pig during the hot weather. In the mere salting of pork it is usual to use only salt and saltpetre. The use of sugar should be avoided in the summer, as its use is likely to result in fermentation in hot weather.

There are two other points in connection with bacon curing on which a change of opinion has taken place, or is taking place. These are the cause of what are called in the trade "seedy bellies," and the effect on the bacon of the female fat pig being in a state of [oe]strum when it is slaughtered. Until quite recently the first of these troubles, and it is a most serious one to the trade, was generally considered to be due to the second. It was believed by curers that the slight inflammation noticeable in the mammary glands of the female pig when she is in heat resulted in these so-called "seedy bellies" if the pig was in that condition when she was slaughtered. This belief may have been either the cause or the result, or both, of the common saying that the meat of a sow pig killed when it was in heat will not take the salt properly, and that it is therefore advisable to wait until this natural condition has passed away before the pig is slaughtered. This contention has been one of the arguments used when the spaying of sow pigs has been advocated. Of late years comparatively few sow pigs have been spayed, so that the unspayed fat pigs have been nearly as numerous as those male pigs which have been castrated, and as the sow pigs come in heat each three weeks, and continue so for from three to five days, a very considerable proportion of them must be in heat when they are slaughtered at the large bacon-curing factories, without any loss resulting. We may, therefore, assume that it matters little whether the pig be in heat or not when it is slaughtered unless the seedy bellies result.

On this point also the verdict is against the common belief, as Messrs. Mackenzie and Marsh have carried out a series of investigations at Cambridge which clearly proved that seedy bellies were equally as common when the sow pigs were not in heat and when they were; but that the discoloration which resembles numbers of small spots of colour varying from dark blue to light red in the mammary glands is merely an excess of pigment, the darker shade being common in pigs with dark coloured hair and skin such as the Large Blacks, Berkshires, etc., and the lighter shade in pigs of the Tamworth breed. In the bacon manufactured from pigs with a white skin and white hair there is no discoloration or seedy bellies.

Although it has been generally considered by bacon curers that pigs of a white colour were preferable for their trade, and this to such an extent that some of the bacon curers in Ireland will pay a slightly higher price for a pig with a white skin, the preference was generally considered to be due to the more presentable appearance of a side of bacon from a white than from a black pig; it would appear that in the future a still greater preference will be observable when it becomes generally known that the bacon made from white pigs is free from seedy bellies.

* * * * *

www.ingramcontent.com/pod-product-compliance
Lightning Source LLC
Chambersburg PA
CBHW070325190526
45169CB00005B/1752